Contents

From the Author 5
From the publisher (and the author) 7
Uncle Sam 11
What is your mission? 15
Stay or Go? 23
Car Talk 35
Gear 43
Firearms 65
When you run out of bullets 75
Tactics and Tips 81
Food 87
Water 99
First-Aid and Medical 103
Know your Zombies 111
Final Thoughts 117
The Navy SEAL Foundation 119

A Navy SEAL's Guide to the Zombie Apocalypse

A Practical Guide for Surviving Anything (even zombies)

Copyright 2022 – Rock Arbor LLC. All rights reserved. The contents of this book, or any part thereof, may not be reprinted or reproduced in any manner whatsoever without written permission from the publisher. This book is for entertainment and general informational purposes, none of which is or should be construed as literal advice or recommendations. This content is unofficial, privately created, and originally written without any oversight, review, or approval by any third party. The publishers of this book claim no sponsorship, endorsement, or partnership of any kind with any third party.

From the Author

Survivalists and preppers know their stuff, but I've experienced some of those life & death scenarios, not just prepared for them or practiced for them. It's a pretty great gig when your 9-to-5 is to put bad guys in their grave, but it comes with some inherent risk. For starters, those bad men tend to not be cooperative with your mission success, so you have to go find them, typically into not very friendly places. To do my job, I've trained and operated in many situations relevant to this book. I've spent combined weeks of my life underwater, hundreds of high-altitude jumps, fought in multiple wars & conflicts both as a SEAL and a military contractor and various alphabet soup agencies, operated in all kinds of climates, and that's all just side effects of carrying out a mission. Staying alive and functioning was merely a condition to completing the objective. Surviving a zombie apocalypse isn't much different. While I can't put you through the training I've done over the past couple decades, I can pass on some of the lessons learned. I've literally dedicated my life to this stuff, and I want to share it with you.

There's some good info in here, but this book is first and foremost for fun. Don't get offended or take anything too seriously.

\\ From the Author

Also, as evidenced in recent years, you apparently get rich when you're a SEAL and write a book. So I figured let's do one about zombies because....well, I wanted to.

And you bought it, so I guess it worked.

Thanks.

From the publisher
(and the author)

"If you're gonna be stupid, you've gotta be tough"

-APOCRYPHAL

"…also, don't sue us."

— THE PUBLISHER OF THIS BOOK

This book is for general information purposes ONLY.

Remember that part on the previous page about "this book is for fun"? Let's expand on that. A lot of the stuff discussed in this book is INSANELY dangerous and in no way, shape, or form are we responsible for any injuries or deaths you cause from attempting *anything* in this book. What do we mean by "*anything*"? Glad you asked. Here's some examples:

From the publisher (and the author)

1. You buy an energy bar suggested in this book and choke on it, that's your fault, not ours. You are solely liable. We accept no responsibility, liability, or fault.

2. You buy a gun and hurt yourself or someone else because your dumbass didn't get proper training (or even if you did get training), that is your fault, not ours. You are solely liable. We accept no responsibility, liability, or fault.

3. If you get a papercut while flipping the pages of this book, that is your fault, not ours. You are solely liable. We accept no responsibility, liability, or fault.

4. You have an allergic reaction to the adhesive on a band-aid for aforementioned papercut, that is your fault, not ours. You are solely liable. We accept no responsibility, liability, or fault.

This book is about self-reliance and survival…in the context of a TOTALLY IMAGINARY SITUATION, so don't run out and start doing stupid things.

Especially for firearms we absolutely recommend that you take beginner, intermediate, and keep going to even advanced training courses. There is no such thing as *'too much training'* for firearms. Courses are available almost anywhere, check out NRAinstructors.org. Everyone has the right to enjoy the 2nd amendment, and shooting guns is just plain awesome, but we believe that right comes with inherent responsibilities. For the love of all that is good, if you've never fired a weapon, TAKE A SAFETY COURSE FIRST BEFORE YOU BUY A GUN. Keep your weapons secured at all times (apocalypse or not), so kids, untrained people, and thieves can't get them. It is YOUR responsibility for every bullet fired from YOUR gun, whether you are the one pulling the trigger or not. Savvy?

This book is written in the hypothetical of an apocalypse, and a zombie apocalypse at that. In other words, the odds of it happening are about as good as Jimmy Hoffa riding a Sasquatch into Times Square with a $400,000,000 winning Lotto ticket. To that point, no one is advocating that in any given emergency situation you do something stupid and rash like loot a store or start shooting guns at anyone who comes within a 1,000 yard of you. Quite the opposite. This book is the worst of worst-case scenarios, so you need to exercise common sense. Not to sound arrogant, but you bought this book because in the context of my expertise and background, you want to learn to think like me, right? Well, I joined the military to <u>help</u> people and fight the fights they cannot. I didn't do it to harm people (except bad guys…bonus!) but those are stories for a different book (btw, make this a #1 best seller and maybe I'll spill some beans). The point is, there are many steps between apocalypse and a mere disaster (ex- earthquake, giant meteor, economic collapse, etc) where you need to use your *skills and knowledge* to HELP OTHERS, not take advantage of them. This book is on the level of mass extinction of the entire human species on a global scale in which the government and social constructs have disappeared a long-time ago. In other words: don't be a dick.

<u>One more time, and say it with me:</u>

I AM RESPONSIBLE FOR ALL OF MY OWN ACTIONS.

Even if it *appears* that we recommend or suggest doing something stupid and you do it, it's still 100% your fault and you cannot sue us.

Ignorance is not a defense here.

Uncle Sam

"I'm from the government and I'm here to help"

-The most terrifying words ever uttered

Uncle Sam is really like that one uncle no one ever invites, but still somehow always shows up to family get-togethers. The creepy one who makes inappropriate jokes, talks way to close while totally unaware of his halitosis, is always asking people to "invest money" in some "big idea" despite his life being one monumental screw-up after another. Then drinks a lot, gets belligerent, and eventually passes out and pees in his pants. The next morning, he blames everyone else for not making him stop drinking earlier, takes all the left-overs and doesn't say thanks as he leaves in a huff.

Now imagine you give that uncle a billion dollars, a laboratory with lots of dangerous chemicals and biological agents, unlimited animal test subjects, and the resources and permission to weaponize one or more of those things. You know what's going happen: your uncle will get drunk, stay up late watching The Incredible Hulk and get some idea of how to make super-soldiers. The next thing you know that moron is injecting himself with chemical cocktail using chimpanzee urine as a base. This example is really fun and I could go on forever, but the point being:

There is a 50,000% chance the zombie outbreak is the government's fault.

Keep this in mind when someone shows up and says "I'm from the government, I'm here to help", what they really mean is "I'm here for the cover-up". This book is to help you with what to do instead of following that pimple-faced 18-year-old Private from the National Guard to the nearest quote-unquote "safe zone". First lesson: whatever secure location they take you to will be complete with high-tensile chain-link fences topped with some nice razor wire. It does just as good of a job of keeping you in just as it keeps the zombies out. The obvious problem is that when, *not* if, the infection gets inside, you're toast.

Still not sure you want to take that advice and decide you'll let the gub'ment care for you instead of being a grown-ass man or grown-ass woman? Here's just a single, but funny, example of government ineptitude for your reading pleasure:

In 1971, some hippies decided they'd break into the FBI to steal documents to help their goals of spreading world peace. You're probably thinking "well, of course that didn't work! A bunch of stinky hippies wouldn't get 2 feet in before they were caught." Unsurprisingly, they weren't successful at picking the locks…so they left a note on the door requesting the door be left unlocked at night.

How stupid can you get?! I mean, they left fingerprints and their handwriting on the *literal door of the FBI*.

Case closed, off to jail, right???

Nope. A government employee of the FBI actually left the doors unlocked the next night as requested. At this point, you're thinking "yeah, but they set a trap, lol!". Negative on that ghost rider. The hippies went in, stole a boatload of classified information like they were browsing their local organic vegetable stand, and left. Yes, really.

Quick side note, considering I might later seek employment by the DOD, CIA, FBI, or any other number of alphabet soup that is our Federal agencies...Love ya Uncle Sam!! All in good fun! JK! LOL! We good right?

What is your mission?

"Social niceties have been suspended. We're in a state of emergency. The world has descended into chaos. Lawlessness and savagery are the order of the day."

- SHELDON, BIG BANG THEORY

Fair warning, this is the boring serious chapter.

Welcome to the new world! No government, no rules, no taxes, real zombie targets everywhere. What's not to love? You have the knowledge. You have the skills. You are prepared and ready to dominate.

But what if you aren't? If you aren't sure, read on.

First order of business, you've got to get your mind right. The knowledge, skills, and physical preparation is only a *result* of having been mentally prepared beforehand. You need to have thought through the potential threats to your safety and health and how to overcome them. Even if all the equipment, guns, food caches magically appeared when you need it, that far from guarantees you'll be OK if

\\ What is your mission?

you don't have the right mindset. This ain't a Hollywood movie. You could be faced with unfathomably difficult decisions that are the difference between life and death. Stuff that you need to steel yourself for so you don't go catatonic when it happens. A lot of people think that the life of a SEAL is non-stop action and excitement and pew-pew-pew'ing. Realistically, 99% of it is train, train, train, then train some more. It is monotonous, boring, and often painful. Nothing about it is glamorous, but the 99% of the time you are training will all keep you alive in that 1% of the time on a mission when the world around you submerges into the fog of war, engulfed in chaos.

This book is written in the context for the medium to long-term survival. The first 24-72 hours I'm not going to cover in much detail because by preparing for the long-term, the short-term will take care of itself for the most part. Let me re-emphasize that point: **If you are prepared for the long-term, the short-term will take care of itself.** Most everyone has some food and water available for a couple days in their house already, preferably at least enough for one week for each person.

While I live in the Midwest in the suburbs of a larger city, every person's situation will be different, and I can't cover them all. If you live in the mountains and hunt your own food, well, you probably aren't reading this book anyway. However, if you live in a place like NYC, you've got a pretty obvious problem when you mix "zombie outbreak" with "high density population". NYC is awesome for a lot of stuff (some of the best steakhouses in the world are there for one), but it is very not awesome for zombies. Point being, I'm not going to talk about either extreme of that spectrum of geography or specifics.

Regardless of where you live though, you have to understand some basics of what will happen in a global scale disaster or apocalypse. Heck, even a regional one, since every zombie outbreak has to start somewhere. Think about those simple things we take for granted, such as how our nation's grocery supply is on a 4-5 day cycle. If zombies are running wild and multiplying rapidly in a concentrated population center and you can't get out, you might have to hole up for a long time. Within hours the looting will start and suddenly there's no more food, or worse: only vegan "meat" products.

While this book is for fun, a serious aspect you need to consider is that people will do a lot to feed their family. How far would you go to feed your child or get medicine to heal your spouse? Personally, I'd lie, cheat, steal, and kill if I had to. To be clear, *there is no nobility to that sentiment.* This is a zero-sum game, and what you take could come at someone else's expense. *You* could be the "bad-guy" to someone else. Do your best to not have to be that person though. If you are stronger, smarter, or more prepared, you not only don't have to be a bad guy, but I would argue that you have a responsibility to do what you can to help others if and when you can. I remind you that I didn't join the military to hurt people weaker than me, I did it to protect them and fight the fights they cannot. Point being, as desperation grows, people will take bigger risks and become more brazen; they have less to lose when the only thing to gain is living to see tomorrow. That leads to bad decisions and mistakes, which leads to getting yourself dead. Don't be *that guy*. If you were smart and prepared beforehand, you likely won't have to consider the need to lie, cheat, steal, or kill. You need to have enough supplies to get through whatever the situation is, or you are screwed from the start. If you can't even get food from the grocery store for the first 24-72 hours, you sure as hell aren't going to get your hands on what you need for the long-term.

Being mentally prepared is not just about having a few non-perishable items in the pantry and a few bottles of water. That's physical preparations: tools, supplies, and equipment. Those inanimate objects will not make you "prepared". The point of this section is whether or

\\ What is your mission?

not you thought about what are willing to do to survive AND who are you willing to die for? What weaknesses do you have that others might exploit? Do you have a plan? *Make one!* What about a back-up plan? *Make one of those too*!

We will get into home defense, preparing, and strategic tactics later, but first let's take a look at the 30,000 ft question. You have to know **"What is my mission?"** That baseline will keep you focused and help you make decisions. Let's use some hypothetical case studies:

Scenario 1: You determine that your mission is to keep your family safe. That's pretty broad, but a pretty damn good mission in my opinion. So let's say you and your family come across a well-guarded stockpile of food and water, and you assess you have a 50/50 chance of forcibly taking it with no injury to yourself or loved ones. The other 50% is you or one of your family members gets hurt or killed. Your mission is "keep your family safe", right? You have to weigh the risk and reward. Do you NEED that food and water, or do you hopefully have your own supplies? If so, the best fight is the one you avoid. Live to fight another day, leave it alone, don't go looking for trouble, and go about your business. However, if you were all starving and that was your *only* choice, then taking that food and water might be the necessary risk because keeping your family safe includes not letting them starve. That's not an easy call. In some form, you will have to hurt someone else to help yourself, even if that harm is merely taking food they need too. You just became someone else's "bad guy", and they have every right to stop you with force. It's a judgment call, maybe they are good folks and you could just ask for help and they'd give it. Be pretty messed up to mow down some innocent people who would have helped you if you'd only asked.

It's unrealistic to think of everyone else being "bad-guys" and that's the only type of person you would encounter. In a vacuum, it's easy to say

"I'd have no problem killing a bad-guy!" but real life doesn't happen in a vacuum. Any "bad guy" becomes very human the instant you point a gun at him. Real talk, some people just won't pull the trigger. They might have thought they were willing, but in that moment, they will freeze and they'll be a deer in headlights. Why? The situation just overloads the circuits in their brain. It's not an easy thing taking a life, even if you are justified, trust me.

Scenario 2: You are traveling alone on foot and break your ankle. Can't call 911. Your immediate situation becomes very simple: Figure it out or you're screwed. That's the harsh reality, and if you haven't thought of that situation until it happens, your odds of "figuring it out" are severely decreased. It's going to hurt like a dog crapping peach seeds but staying there to avoid the pain won't get you out of the problem. It is imperative that you be able to differentiate pain from injury. You want to mitigate long-term injury, but also need to stay alive to see tomorrow. As for physical preparations, even just some duct tape and zip ties are better than nothing. Anything can work as gauze to stop bleeding, just tear some fabric off your shirt, duct-tape it, and keep moving. That stuff only helps if you already have it with you though.

Here's a great real-life example from an American hero, Marcus Luttrel, author of "Lone Survivor". If you don't know him, the 30 second condensed version is he is a SEAL who was shot multiple times in a hostile foreign country surrounded by the enemy who was hell-bent on killing him, and his entire team was already killed. None of his gunshot wounds were *immediately* fatal (obviously), but he would be dead in hours, maybe days if he was "lucky". SEAL missions require moving quickly and quietly, so you aren't loaded with gear and medical gear is limited to triage which is enough to keep you alive for a medevac, which wasn't coming anyway in his case. So he literally shoved mud into his bullet holes to slow the bleeding. MUD! After mitigating damage to his body as best he could, the next order of business was to get somewhere safer. In this case, he was so badly injured he could only drag himself an arm's reach at a time...so that's what he did. Reach out, grab some dirt, and pull himself a few inches. Repeat. He did that until he reached a village, who fortunately provided helped hide him.

Scenario 3: You are traveling with a group and someone is badly injured. This is callous as hell, but the first thing you have to do is decide how much time and effort you can afford to put into this person. Told you it was calloused. That's not a purely objective decision of "how bad is the injury", there is a human element of "who is it that's injured"? Let's think about the injury component first. If it's a broken bone, sling it, bind it, and keep going. But if it's a gunshot wound or arterial bleed, you could stop and do your best for them, but to what end? If they only have minutes, hours, or days left without advanced medical care, are you willing to waste consumable medical supplies to delay the inevitable? In the heat of the moment, your human emotional instinct to help could override the objective reality, but a few days later when the emotion has subsided, you could regret it big time. I'm sorry to disappoint you if you're hoping that I'll tell you universally what the right answer is here. There isn't one.

The next component is "who is injured?". Let's say carrying or dragging them isn't a viable option for whatever reason. You have to make the call: they'll take up resources, slow us down, and probably not live through it anyway. Sounds logical, but actually making and living with a decision to leave someone is pretty damned rough. Not making a decision is in fact a decision itself also. But I'll use myself as an example: my mission is "keep my family safe", and it if was my wife or kid, I'd say "Screw it. I'm going to stay here, I'll use up my supplies. consequences be damned." So again: What is your mission? In my opinion, staying alive is for the purpose of living my life, and the question changes from "What you are willing to do to stay alive" to "who are you willing to die for". Dropping some deep philosophical stuff on y'all.

TL/DR:

You'll need to become self-reliant, train your brain to think outside the box, and dig deep inside to get tougher than you've ever been. Even the best decision has a price, so try to do as little harm as possible.

Your humanity and self-respect are worth a lot, but only you can decide that value. You'll have to make compromises, there are no easy decisions…well, except if you come across a cache of beer and bourbon, take it…but remember to do the best you can to live a life you don't regret, so leave a little for the poor guy you took it from. You have to live long enough to get there, but the collective goal for everyone is to eventually rebuild society. It could be years, decades, or generations. So help those you can help, avoid hurting those you don't have to hurt, and hope that others return the favor.

Stay or Go?

plans are useless, but planning is indispensable.

– Dwight D. Eisenhower

You either made it through the "mental toughness" speech, or you skipped ahead. It's OK, don't blame you. It was boring enough to write, I can't imagine reading it. If you did flip right by it, here's the generic summary – shape up or ship out, let's see what you're made of, and never quit! You can do it!

Next up is figuring out whether to hunker down or get out of Dodge. You should have a plan for both. This is the first choice you'll have to make assuming you're the leader… and you definitely want to be the leader. Being a leader in the apocalypse isn't about power or respect, it's about making decisions for yourself and, if any of them survive, the ones you love. Having family or friends alive can be a double-edged sword. Of course, it's great that your loved ones didn't met a gruesome death at the hands of decaying soul-suckers, but holy burden and godforsaken emotional attachments…what's wrong with these morons?!?!

Stay or Go?

If you think your husband is pathetic when he's gets drunk watching football every Sunday, imagine him being drunk every Sunday when it's his turn to watch for zombies. "Sorry, honey, I had a few too many and needed a nap," isn't as forgivable when you have to send your now zombified children to the hereafter with a couple of shotgun blasts.

Does your wife nag you now when you don't take your shoes off after walking in the house because you're tracking dirt? How do you think she's going to feel when you haven't showered for weeks and start eating racoons? Will she approve when you go out hunting with the boys?

Don't even get me started on the children. Tiny, little bad decision makers. When it comes to saving family though, it is the one time to look beyond self-preservation and make irrational, emotional-based decisions. That being said… family is everything. If there's no love left in the world, then what are we trying to survive for anyway? Always fight to keep your annoying family alive.

Back to leadership. Out of all the things you barter, accept, or surrender, never let it be your freedom. You are only as safe as you can keep yourself.

Remember, we're assuming the apocalypse is full swing at this point. It's survival of the fittest in the most real sense. You'll see the best and worst of human nature emerge.

To determine your mission, you need to answer some basic questions of whether staying home is a viable option:

If you want to stay:

- Can I defend my location?
- How so and with what?
- Against what kind of threat?

- For how long?
- Is it a temporary, or do I want to invest time and resources to make it permanent?

If you need to bail out:

- Where are you going?
- Do you have supplies to last long enough to get there?
- Do you know the terrain?
- What other risks you are incurring by going?

Guess what? You don't need a plan, because "a" plan indicates singular. You need "a plan" for each scenario, and every plan must remain fluid and adaptable to an ever-changing situation.

Most people would want to stay in their home, me included. Man's law or nature's law, it's your home, it's your property, you don't want to abandon it. From a practical standpoint, living in Wyoming, Colorado, Texas or areas that have remote cities would be ideal for getting to stay in your home. The town/city itself may have a few thousand or even tens of thousands of people living there, but it's far enough from the neighboring cities to be surrounded by large areas of empty land. Even the building structures and city plans typically have larger and more spaced out lots, so you aren't crammed in with lots of weak points to consider.

One small town factor you are just going to have to buck up and get over is that you will likely know everyone that is trying to kill you. Everyone in the nearby vicinity that has turned into a zombie was once someone you knew. The first guy that's going to try to eat your brains is your good ol' next door neighbor. You've let him borrow your lawnmower, grilled him a steak, and even shared plenty of cold beers with him, but now there's something else he wants: your brains.

Stay or Go?

This is non-negotiable. Kill him. The same goes for sweet old ladies, your kids' pediatrician, and your local girl scout troop.

In contrast, hopefully you don't live in "the big city." Objectively it's a strategical nightmare, but also, you have a higher than average number of crazies and idiots. Think about it, if you escaped from an insane asylum where would you go, New York City or a small town in Kansas? Hands down, you would go blend your delirium into the city that never sleeps. Crazy is open 24/7 in NYC.

What will happen during the Apocalypse, is a lot of people naturally won't want to leave, even if it's a terribly stupid decision. They aren't thinking strategically and have a false sense of security by staying in a familiar place. They're not willing to leave Orange County, California and go to Montana, because they've never been there, have no idea where they'd go specifically, and they're scared. They feel safe in their condo or apartment, even though they have no food or weapons. Collectively, the city will go into panic mode, then all the looting and shooting starts, and it turns into a scene out of Mad Max. Not to mention that there is a likely chance that a big population center would be ground zero of the outbreak.

If you think you have a better chance of being rescued from a big city or somebody magically showing up with a cure, then you are horribly mistaken. I don't know the locations of the top-secret military bunkers that will be working on the cure - to benefit themselves and only themselves – but I can promise you there isn't one in Atlanta or Minneapolis. No one is coming for you. Those first responders are busy keeping themselves and their families alive.

No matter what precipitates a post-apocalyptic world, you'll be in a much better position if you at least think of *something*.

What would I PERSONALLY do if I had to leave?

Keep in mind I live in the Midwest, no major fault lines or anything to destroy the roads or radically shift the terrain I'll encounter. I'd ideally throw my stuff in my jeep and head westward because I have family there and would want to get to them. I don't know how far I'll get, but that would be my baseline plan. If traveling by vehicle becomes a non-option, I go on foot. And by non-option, I very much mean not an option at all. If I have to steal, kill, or barter.... If traveling via the roads, through cornfields, or down rivers is a possibility, that's where you will find me. Why is a car so valuable?

You can go further distances in a car than you can on foot. I know… all these years as a Navy SEAL and I have that rare knowledge to pass on to you. You're welcome. Also, a car can provide you with a barrier. It can replace someone on lookout if you're traveling alone. If you stop to rest (which unfortunately you will have to do from time to time if you're human), you can lock the car and sleep. This is invaluable if you're traveling alone. Sure, you're going to wake up to a hoard of zombies clawing outside your car, but this is much better than waking up to a hoard of zombies clawing on your face. Hit the gas and go.

Another great non-human asset to provide lookout for you is a dog. If you don't have a dog and you must steal one, pick a breed that's not known for its loyalty like a lab or golden retriever. Any lab will be happy to see you for a tennis ball and a belly rub. I don't recommend stealing a dog like a German Shepard, Pit bull, or Rottweiler. You can team up with a Chihuahua in a pinch, at least you know it will yap at any sign of danger… including a leaf blowing in the wind. It's better than nothing. I'm not so concerned with an "attack" dog as I am a breathing alarm system.

Stay or Go?

So, worst case scenario... Me and my newly-trusted Chihuahua are on foot and heading west. I've already got the appropriate clothes on, and I have my go-bag ready to throw on my back. Anything that is too heavy to carry gets left behind, depending on how far I have to travel. The next objective is to conserve my food and water, find more food and water, and keep injury avoidance as a top priority.

As for traveling, there are a lot of "it depends" for best strategy. Generally speaking though, be sure you know where water is. It takes weeks to die of starvation, but only days of dehydration, and depending on the climate and time of year coupled with energy consuming terrain, you will need significant hydration. If you are lucky enough to have a river, stick close to it.

Regarding your route, try to balance concealment with safety. Remember, it's about mitigating risks, which often required a compromise. Think about it, you don't want to make yourself exposed because that could make you vulnerable to an ambush. However, you'd be foolish to attempt overly difficult terrain that would likely result in injury, consuming too much energy, or other personal safety factors. Using the example of being lucky enough to have a river for water, I'd personally blaze a new trail through the trees (not on an actual premade trail by the Forest Service), just close enough to hear the river. Now I've covered my bases: I have a source to replenish my water, I've got decent concealment by traveling a path that wouldn't be watched by potential bad-guys, and it's minimal risk for injury. To boot, a river is a great for navigational purposes.

What is a defensible location?

A lot of people think of their home as being very safe. It's not. Every window and door that's an exit point for you, is an entry point for somebody else. Homes are designed to keep the weather outside, not zombies outside. You have to physically stop someone from coming in, and glass windows and standard residential doors will barely slow someone down. To fix this, you'd need truckloads of heavy duty plywood, 2x4's, and structural grade anchors and bolts to truly "stop" someone from gaining access. Even then, it would take you a week or longer if you know what you're doing and cost you thousands of dollars, and after all that work it's still flammable and sure as hell not bullet proof. Why does bullet proof matter? Because humans are just as dangerous, if not more, than zombies. You could turn your home into a safe zone, but trust me… It's easier to get to Fort Knox than it is to build Fort Knox.

This is a good time to discuss Cover vs. Concealment.

- **Concealment** is just hiding behind a bush. Bad guys can't see you, but it won't stop their bullets.
- **Cover** on the other hand is something that will stop a bullet. Think of bullet proof glass. Even though they can see you, their bullets won't get you.

In the case of a house, securing it to physically stop someone is along the lines of "cover," but that's a difficult task to be truly effective. However, concealment is more easily accomplished and could be just as useful. Cover up the windows however needed. Use paper and tape, paint it over, whatever, so that people can't see in and can't see light during the night. Do what you can to make it look run down so that it doesn't look inviting to others. That might mean vandalizing your own property. Make it look like it's been looted and ransacked and no reason to come in, but don't overdo it compared to your neighbors or else it has the reverse effect of standing out. And, above all else, no upkeep. If you have a partner who insists on mowing the lawn, trimming the hedges, or watering the azaleas then you have to leave them immediately. Everything I said earlier about family being everything doesn't apply to you. Take the kids though, they deserve a fighting chance.

A quick note about concealment vs cover: This is for hiding from the looters, outlaws, murderers, etc. Zombies are good at one thing and one thing only: persistently tracking humans in order to consumer their flesh. They are (usually, not always) slow and weak, but they aren't your five-year old cousin that sucks at hide and seek; they will find you under a pile of leaves.

Next up for what constitutes a defensible location is that high ground is always desirable. If you leave your house and decide to set up a potentially permanent or long-term base-camp, something high up on a hill with trees for concealment is great, especially if there is enough clearing around the immediate camp so you can't be ambushed. "High ground" is relative to a few dozen meters radius, not hiding up in a tall tree like a squirrel. A mile radius or more around you would be ideal with alarms set on the perimeter. Another important benefit of high ground is how it affects whatever human waste disposal system you have set up. If you can't figure out how that might be beneficial, just imagine if you chose low ground instead… It's hard to dump everyone's dumps when you're living in a valley.

You must also consider the size of what you want to defend. Is it just you? Just you and some buddies? You and your kids? If you've got some other shooters, that's good, that's fewer people to defend and more people that are helping defend. A bigger space means more area that has to be defended and potentially more weak points. If I'm alone, I can defend my car, that's all I need. If I've got a dozen guys who can shoot, we can have a way stouter location with bigger and stronger defenses set up. The need to escape and evade danger is replaced with focusing on the best defense possible.

Considerations for traveling:

Take care of your gear and your body

Here's a tongue twister: Take care of yourself by taking care of what takes care of you. Something non-military folks might not realize is that no matter what you've just been through, no matter how tired or hungry you are, unless you've got an injury that needs immediate attention, the first and most important thing is to make sure your weapon is clean and ready to fight. Even if you've reached what you consider relative safety, you never know when a threat could pop up and you'll need an operational weapon. It does no good if your belly is full but your weapon is empty when you get ambushed. You can go weeks without food and a few days without water, but not so well with a bullet through in your noggin.

Especially if you're traveling by foot, the next most important priority is your feet. Take your socks off and put dry ones on regularly. If you can't change socks, still air your feet out. Don't just sit there in wet nasty socks if you don't have to. It's a delicate balance, but you have to take care of your feet and still be ready to move again in a moment's notice. Once everything else is done and squared away so that you're ready any second for fight or flight, then tend to your triaged injuries and get some food and water.

Exposure

The next fastest thing to kill you is the environment. If it is freezing, try to get burrowed down. Pine needles, leaves, dirt, anything to help insulate you. You can die from hypothermia in hours, so a sudden cold front or snowstorm can be lethal, and even good cold-weather clothes aren't necessarily meant for long-term exposure measured in days or weeks. Try to stay dry as well because you lose body-heat exponentially more rapidly when you are wet. Yes, sharing and conserving body heat is an effective way to stay warm. I don't care what you say, real men snuggle with other men when your life depends on it. If it makes you feel emasculated, just fist fight over who gets to be the big spoon.

Land Navigation

Let's talk about land navigation. We will cover this more in the chapter about gear, but Siri isn't going to give you directions anymore. You will need to already have a decent map and know how to use it. Of course, this is assuming you have a longer distance to go, but it is something you should have on hand regardless because survival situations can change rapidly, and you might need to be on the move anytime. The roads and highways may not have cars driving on them anymore, but those roads won't just disappear, so use that to your advantage when planning routes. That said, you may need to use caution to avoid walking into any ambushes; I'd achieve this by using the highway as a marker, but not necessarily walking along it.

Another option is to use rivers, something my grandpa once referred to as God's highway. I touched on that earlier, but if you get lost, you can follow any river back to civilization in a pinch.

Blend in

This isn't just blending in as in wearing camouflage. Think about your tactics and strategy when out in "public." This is practical whether you're staying put or traveling. Don't shave. Don't brush your hair. Don't wear nice clothes. No one will be intimidated by you parading around in expensive survival clothes looking all neat and clean. You'll just put a target on your back and someone will be willing to take the risk to rob you of all that nice gear you've got.

Also, blending in does not mean to act like a zombie. You will not fool the zombies and if you are able to fool the humans, there is a good chance they will kill you. So, stop moaning and dragging your feet. Simply walk around like a hobo minding his own business and hopefully no one will bother you.

In fact, let's back up and let me give you a little practical advice that took me multiple missions to fully grasp. You have to know your enemy. Whether it's your mother-in-law, the popular girls in middle school, or national terrorist threats, you have to be aware of their strengths and weaknesses. I can only speculate on the traits of your particular zombies, but I'm guessing they hunt by smell. Pay attention to what seems to draw them to you. Do zombies show up when you're in a large group, when someone is injured, after bathing, after a long period of not bathing, after human waste piles up, etc...

Okay, back to business... If you want to hide from zombie among zombies (not recommended) then you must figure out what smells attract them and block them with smells that don't.

TLDR: Concealment hides you, but cover stops bullets. Consider both when setting up a location. If you have to travel, know where you are going, have a plan, and try not to attract attention to yourself. Don't forget the basics, such as the fact that cold weather can kill you in a few hours if you aren't prepared, and try to know where your next source of water is before leaving that one you have.

CAR TALK

"When I die, I want to go peacefully in my sleep like my grandfather. Not screaming in terror like his passengers."

There are no specific rules on who will survive a zombie apocalypse, but I can tell you there are definitely some trends. As mentioned, a good hunk of the military is going to survive (they were probably the ones that caused it anyway). Which means you are left with a bevy of scientists to create a vaccine, doctors to look for a cure, and engineers to rebuild our once great cities. So, the good news is all you have to do is survive for a handful of decades while they get that all sorted out. The high military survival rate also means you have plenty of trained killers available, cross your fingers real tight and hope that's a good thing.

Who else survives? Prepared hillbillies, smart cops, tough firemen, and a whole bunch unwittingly lucky folks like accountants, middle management, and stoned fast food employees who will be worthless in a zombie apocalypse. People who you could describe as "bumbling" will inevitably survive the initial blast. It defies logic, it's not fair, but just how things go.

Now let's talk about who never survives that we could actually use… the blue-collar heroes your grandfather would consider to be a "man's man." There will be no plumbers, electricians, or mechanics available in the yellow pages. For better or worse, anywhere you damn well please can become a toilet in the Apocalypse, so we can live without plumbers. Electricity will be one of the first major societal functions to collapse, so there's nothing we can do there.

Cars, on the other hand, are something we will definitely need. You've got to learn basic mechanical skills.

Here are the four most essential post-Apocalypse car skills, in order of priority:

1. How to siphon gas
2. How to change a flat tire
3. How to change the oil
4. How to hot wire a car

How to siphon gas

Whether gas stations are not working, not nearby, or surrounded by brain-craving flesh wads, there are going to be times where siphoning gas is the better option. The easy way? You have a pump siphon on you or found one (it's worth buying now, they are about $20 bucks online). If you're in a pinch, keep reading. You will need a hose, a gas can, a working human mouth and lungs, and, if you're targeting newer models (avoid if possible), some type of stick or screwdriver to break open the gas tank cap. If you don't know where to find a human mouth or a stick, I'm sorry to say this book won't be able to help you.

In fact, you have already long outlived your potential. Congratulations for that and sorry for your impending death by the zombie hoard.

Where do you find the hose? Lucky for you, cars are full of hoses. Pick any as long as it's long enough to reach from all the way inside the tank of the car, coming up through the nozzle hole, and down to your container on the ground; this is a physics thing, I'm not getting into it if you don't understand. And, just in case you are one of those "bumbling" survivors mentioned above… whatever hose you choose, take it from a car that you are <u>NOT</u> planning on driving. The hose comes from the car whose gas you are siphoning, not driving.

Where to find a gas can? That is one of those rarely discussed zombie survival necessities that you will just need to have on hand. If the gas stations are fully looted, you will have to try the deceased garages and sheds. And yes, there are zombies in the shed. Weapons up. There are *always* zombies in the shed; it's just one of those weird things that right before people turn into zombies, they jump into sheds and other inconvenient spaces for no apparent reason. Anyway, now you are probably down two survivors, but up one gas can, let's move on.

Now, approach the car you are trying to siphon gas from. Remove the gas cap, using your stick or screwdriver if needed to pry open so you can slide your hose in. If you go nice and slow, you should be able to feel when the hose has reached the gasoline, but we're past the point of safety concerns, so just jam it in there. Once the hose is in, put your mouth around the outside of the hose and suck. As soon as the gasoline hits your mouth transfer the hose to the gas can on the ground. Now, the age-old argument of spit vs. swallow. Considering I'm writing this book to help you survive, I highly recommend spitting out any gas and rinsing with copious amounts of fresh water. Break water rations, it's worth it. On the other hand, drinking a small amount of gasoline does create a cheap, albeit painful and risky level of intoxication. If you've reached the "ah, screw it" mentality that so many do when you spend most of your time fighting zombies and living like a hobo on the run, you may make the conscious decision to become the group liability. There are a few precautions to take when drinking gasoline. First and foremost, a little goes a long way and a lot can kill you. Don't go all college pledge-class on me. Definitely don't enjoy a smoke for what should be obvious reasons (ie - you exploding). You know what, just spit it out. I'll teach you how to make hooch in a later chapter.

Next, take the gas from the gas can and place into the car you plan on driving. If you ignored my warning and are drunk on 87 octane, let someone else drive. Zombie apocalypse or not, drunk driving isn't cool.

Changing the Oil

This is a relatively simple thing to do for most weekend warrior types, but if you've never changed the oil on your car it's a 3 step process in the normal world.

1. Unscrew the plug on the oil pan and let the old oil go into a pan.
    a. This is underneath the engine compartment and looks like a flat-bottomed metal tank (like a gas tank). If you've never done this…it's towards the front end of the car.

2. Replace the oil filter and screw the plug back in.
    a. The filter is usually a white metal "can" shaped thing accessible usually from under the car, also located in the engine compartment.

3. Pour in new oil.
    a. This is poured in from the top of the engine compartment. The cap is marked with an engine oil symbol (duh).

However, this isn't the normal world, and you'll need to make some compromises possibly. First, the EPA will hate me for saying this, but recycling old oil is a pretty low priority given your present circumstances. You always want to park on a flat surface, but normally you'd want to be sure that you don't spill used oil and if it does spill, it's not going to flow into sewers. In the apocalypse however, wherever your car happens to be when you need to change the oil is the right place, but you need to keep in mind that after a few liters of dirty black oil is dumped on the ground, you have to get back under there to put the plug back in. That means you'll still want some way to catch it or do it over grass or dirt to absorb it (sorry, EPA). Possibly just park on a slight incline and crawl under from the higher end, letting the oil run downhill away from you.

Next, you'll want to use the right kind of oil, which can easily be found in the manual or sometimes on the oil cap. The obvious problem is you may not have the exact right kind, but it won't ruin your engine to use a different weight oil than recommended. For those not in the know, "weight" oil refers to those numbers you see: 10w20, 10w30, and so on. Those numbers refer to the viscosity of the oil at a cold temperature, and then the viscosity when the engine is warm. Without getting into the physics and science of it, the different types are meant for warmer or colder climates in various engines. Unless you are in sub-zero temperatures, your car will be better off with clean oil than dirty oil or no oil at all, so grab whatever is available in a worst-case scenario.

The other issue is the filter. You might be hard pressed to find the right one, in which case you'll have to reuse it. This is a terrible long-term issue, but it's better than nothing. I'd suggest taking that old filter and turning it upside down to drain out as much old as you can. Then, flip it right-side up and fill it up with clean oil and let it sit for a while. Then, repeat the process by dumping out that oil. This is far from perfect, but will at least dilute and remove some of the nasty crud in that filter and slightly improve its performance.

How to change a flat tire

If you are lucky enough that your pop is still alive, thank him if you already know how to change a tire. If he didn't teach you and he's still alive, give him a solid punch in the shoulder for failing that checklist item of fatherhood.

As always, electronics are history so no way to YouTube "how to change a tire", but odds are the owner's manual will have some instructions. Here's the basics:

1. First, if you think you blew a tire, don't keep driving. It will destroy the rim of the wheel, and quite possibly other parts of your car. The only exception is life or death and you have no choice but to keep driving, regardless of damage to the car.

2. Park on a flat surface. (By now, it goes without saying to find a nice wide-open area to avoid zombies). If you don't have access to flat surface, park such that the car is not facing uphill or downhill, but laterally such that the flat tire is on the uphill side. For those unfamiliar with basic physics, it won't be easy to lift the car from the low-side.

3. Put the car in Park and engage the emergency parking brake and get out the spare tire, jack, and any other equipment.

4. Place jack under the chassis of the car, or in the designated spot as per the owner's manual. Put spare tire under the car, under the chassis; If the jack fails, it will keep the car from smashing into the ground and possibly destroying parts you cannot fix. Don't start jacking the car up yet.

5. Before jacking the car up, you must loosen the lug-nuts. It sometimes takes some serious force to break them loose, so you'll need the weight of the car to keep the wheel from spinning as you try to spin the lugs partially loose.

6. Once the lug-nuts are *partially* loosened, you can jack up the car. Only jack it up a couple inches of clearance, you don't need more than that.

7. Completely remove the lug-nuts, placing them in your pocket or somewhere you will not lose them!

8. Set the old tire under the chassis like you did the spare, THEN pull the spare out. Do NOT lay under the car with a jack holding it up! EVER!!!

9. Fit the old tire on and put on all the lug-nuts as tight as you can using only your fingers. You might have to push on the tire at the top and sides to wiggle it around as you tighten the lugs.

10. Once the new tire is on and lug-nuts are snug such that it doesn't "wiggle", lower the car onto the ground.

11. The last step is to fully tighten the lug-nuts back down with the lug wrench. It is critical that you go in an alternating "star" pattern, and not just to each lug in order in a circle. That can result in the tire not being perfectly centered, causing a potentially dangerous situation and damage your vehicle. If you don't understand, remember how literally everyone draws a star? You start with one lug, then do the lug on the far side, then back again, and so on.

\\ CAR TALK

How to steal a car

I got good news and bad news. Bad news first: real life is not like the movies and you cannot just rip a bunch of wires from under the dash and start a car. There is no such thing as a generalized magic of "touch the red wire to blue wire, then pump the gas twice". You would essentially need to know the exact wires and the exact process for each and every make and model of car on earth for this to be effective.

Good news: some cars and trucks are easier to steal than you might think. Older vehicles without keyless entry are the way to go if you need some wheels in the apocalypse. First, they likely won't have an alarm, and everything is mechanical, not computerized. Generally, a flat head screwdriver and a hammer are the only tools you'll need.

First, you need to get access to the vehicle. Ideally, you want to keep the glass intact, so try using the flat head screwdriver to rip the door handle cover off and see if you can manually pull a lever to unlock the door. If you have a "slim jim" on you, and you know halfway how to use it, you can try that…if you don't know what a slim jim is, don't worry about it, this doesn't apply to you. Your last option is to break a window. Try to use only the minimal amount of force necessary to break the glass, and be smart and go with a back window so you don't get glass shards in your butt-cheeks from the drivers seat.

Once in the car, jam the screw driver into the ignition and give it a couple solid thumps with the hammer. This will destroy the ignition cylinder for any future use with the key, but will often allow you to start the car.

Not exactly sexy and glamorous like a Hollywood spy flick, but brute force is the name of the game here in the real world of car theft.

GEAR

Beans, Bullets, and Band-aids.

Practical not Tacti-cool

I need to talk about something first before we get into the specific gear lists. When you imagine a SEAL going out to survive the zombie apocalypse, what do you imagine he looks like? Decked out with a full-kit, looking Operator AF?

NO. Just, no. SEALs and other Special Operations groups try not to stand out. In many cases, they try to blend in. The beards that operators are somewhat famous for in recent years isn't a trend or fashion statement, we work a lot in the middle east where men normally don't shave their faces. It's not to be "cool", it's practical for our job, not to attract attention. You might be surprised what a SEAL would take and how he would dress. There is a picture of me and my team in a literal warzone in Iraq, and if I took off my holster and plate carrier, I look like I could be strolling through a local grocery store. I'm wearing a breathable polo shirt, cargo pants, and lace up boots with a ball cap. I don't need camo or a uniform, so I'm not wearing it. Simple as that. Camo is to blend into your surroundings, and those clothes made me blend in in that context.

I am not saying camouflage is stupid and has no purpose. Rather, treat it as any other tool or equipment. For example, your down-stuffed winter jacket that can handle -50 degree temperature is a fine thing in a blizzard, but you won't wear it in July in Texas would you? Same thing with camo. Have it, but use it when you need it.

Skipping ahead to weapons a bit, but still in terms of tacti-cool, a tricked-out AR-15 platform is a good weapon, but that's for warfighting. Don't' get me wrong, I love them and I have a few. Maybe that's what you need if you have to fight your way through crowds of zombies. It could be argued that if you're smart though, you'd have already left before you get overrun with brain-eaters. We'll get more into that later, but point being, don't try to be like a SEAL like you see on TV. It's just not real. Look like a "regular guy". Become a physical embodiment of background noise. Attracting attention or looking like a "tough guy" isn't going to help you and doesn't intimidate zombies (or anyone else for that matter).

Some "tacti-cool" items have a basis in practicality, just keep it in context and don't parade around. A good example is a Shemag. You may not know the word, but you all know what it looks like: it's the middle-eastern "scarf" that people tie around their face/neck. These things are very lightweight, but kind of bunch up in layers when you wear it, which can help keep you cool in scorching heat and protect your face and neck from the sun, but also helps hold in heat during the cold. Although thin, it's a lot of material and can be used in a pinch for medical aid such as a bandage or tourniquet. Lastly, it can provide some concealment since it's typically camouflage colors. Don't get carried away and wear it everywhere like a D-bag though. You might as well put on skinny jeans too.

GEAR LIST

Let's keep in mind that this stuff is a mix of short-term to long-term gear you'll need. You need to exercise common sense and be realistic about consumable products. For example, you need food, but 2 granola bars is obviously a very short-term help. Equipment, such as a knife, you want to invest in because that needs to stay a long-term piece of gear and you can't replace it easily.

Clothes

Seems basic, but worth covering, just in case. Dress for the elements but remember that seasons change, and you need contingency plans.

- **Pants:** Jeans are a no-go. They are great for working outdoors and for hunting and camping trips, but not medium to long-term survival when you don't have a home and washing machine to go to. Because they are made of cotton, jeans hold water making them very heavy, cause absolute loss of insulation, and take forever to dry out. Instead, get something meant for the purpose of being outdoors. 5.11 is a great brand, but you can find similar option that are store-label stuff at REI, Cabelas, or other outdoor stores for possibly cheaper. Look for quality constructed materials that feel strong and rugged such as having heavy grade metal zippers, solid metal clasps and buttons, etc. Plastic zippers and such are an indicator of cheap quality. I'd go for something that is water resistant as well, and preferably has reinforced knees if you can find it. No need for camo, but a regular solid green, brown, or tan can accomplish the same thing.

- **Shirt:** Again, avoid cotton. I would suggest a long-sleeve button up shirt made mostly from nylon and poly-blend or wool. They are very light-weight and comfortable in 100 degrees down to mid-60's. It keeps the sun and light-breezes off your skin equally well, doesn't hold water so you dry out faster, and are generally pretty durable. I would suggest having two shirts, one on your body and one in your pack.

- **Jacket:** This is a trickier one to nail down because it depends on where you are. If you live in South Texas, you'll never see temperatures below 20-30 degrees except maybe a couple times a year when a cold-front blows in, and it only lasts for a day or so. If you live in the mid-west, you could see *high's* in the 20's for weeks on end. My suggestion is that wherever you live, plan on ONE good coat. You can't lug around multiple options, that space in your pack needs to be reserved for food and supplies, not creature comforts. Try to get maximum thermal protection, but avoid bulky stuffing. A good jacket should be an investment, so don't skimp. It is a definite bonus if the jacket is water-proof or water-resistant too. Although I said "only one jacket", I give an exception to coats that have liners that can be removed and used separately. Use your best judgement here.

- **Cold-Weather considerations:** Base layers: In extremely cold weather, layers are your friend. Along with your usual shirt as listed above, you need some thermals or "long-johns" or "long underwear". Technically, called "base layers". Usually they are available in light, medium, or heavy-weight materials, which is relative to how cold of temperatures you need to endure. Generally, as long as it is made from wool or polyester, you're good to go.No reason to spend money on name-brand stuff here. Anything in the way of $15-40 is just fine, with the caveat that if you live in extremely cold climates, then you might want to splurge and get top of the line stuff and maybe some extra layers as well.

- **Gloves:** Again, it depends on where you live, but having some lighter-weight gloves as well as heavy-duty mittens is a good call in my opinion. Your extremities get coldest the fastest, specifically your fingers. You don't want to lose dexterity in your hands and fingers. If you only have thinner gloves, they won't keep you warm enough for long. If you only have heavier gloves or mittens, you might need to take them off to use your fingers in which case you have zero insulation for a while. I would suggest looking at hunting outfitters for some insulated hunting or work gloves. These are made for hunters going after large game in the North and are great for survival purposes too. For lightweight gloves, any running or fitness brand would do the trick. Get something with polyester or wool, but nice and thin; just enough to keep the nip of the cold breeze off your fingers. Also, try the gloves on! Of all the dumb things to mess you up, don't let one of them be gloves you ordered in the mail and assumed fit. Also, avoid 'ski-gloves'. They are great for keeping your hands warm, but terrible for survival use because they are not designed to be rugged or stand-up to abuse such as climbing rocks and using tools. They are made for holding ski-poles, and that's it.

- **Hats, etc:** It is actually an old wives' tale that you lose most heat through your head, but, you still want to keep your brain bucket warm. I'm not going to make a recommendation here for brands, just materials. You guessed it: no cotton, stick with wool or poly-blend. Go for well-constructed and densely stitched, but as thin as possible. Any cheap ski cap will work technically, but they are bulkier and less effective. That's a double-whammy. I highly suggest getting one that pulls down over your face; you'll thank me if there are windy conditions.

- **Glasses:** Invest in some decent sun-glasses. This is one of those things that people will overlook as a 'necessary' item for survival, but I consider highly important. You need to protect your vision; you only have two eyes and no replacements. The sunglasses help protect against mild-debris and such that could otherwise permanently damage your eyes, but also keep your eyes from getting strained and fatigued from the bright sun. I have no allegiance to any one brand, but something name-brand is a good option here. I like Wiley X glasses, they are medium-priced at around $60-100 on average, as opposed to other brands that can be up to $300. They are American made, designed for the military-specs, and just personal preference of comfort and fit. Don't buy the $10-25 knock-offs by the register at the grocery store. Polarization is a personal call, it costs a little extra, but I would suggest you spend the extra few bucks.

- **Socks:** An often-overlooked item of importance that can wreck a mission is socks. As I mentioned once already, if you are going somewhere on foot, your feet are one of your most valuable assets and you damn well better take care of them. Do not ever wear cotton socks. Remember, when it comes to survival gear, "cotton is rotten". Cotton has many uses, but out in the wild, it holds water and can cause blisters faster than any other material. Get some real outdoor hiking socks made from wool or some other material made for the purpose of hiking or outdoors. Bring extra pairs! At least 2 extras on top of what is already on your feet. You can go without underwear and your shirt can turn stiff from dirt and sweat, but you must have good socks. Also, whenever you stop and have the opportunity, rinse your dirty socks if at all possible. Get all the dirt and grime out that you can and let them dry so you have a constant rotation available because you never know when you'll get the chance later.

Holsters

I know that drop-leg holsters are definitely considered "tacti-cool", but they can serve a legitimate purpose. I debated that this should be in the "weapons" section, but decided it deserves to be kept broad stroke here amongst other "gear". First, let's be clear that a good retention holster on your hip is the best option. It must be very secure to your body so it won't allow your weapon to accidentally fall out and is the best option for the fastest draw. However, if you're carrying a big pack with hip/waist support, that holster on your belt will pose an issue and become inaccessible. In that case, dropping your handgun down on your thigh will fix that problem with minimal drawback. A little lower than ideal for quick-draw because you have to reach a bit further than your arm naturally wants to go, but practice can mitigate that problem. On that note, remember how I said SEAL's spend most of their time doing training, training, and more training? Don't take anything for granted, you'd be amazed how many civilians and even police officers snag their weapon on something while drawing because of not enough practice and familiarity. Like in sports, practice how you want to play: if you only practice handgun skills on a non-moving target in an air-conditioned range, always at the same distance and give yourself plenty of time to line-up each shot, and never fire from a draw, that won't help one bit in the real world. In contrast, SEALs know every thread and button and clip of everything they carry from countless hours of *practice*, and they practice in simulated real-world scenarios. The downside is most ranges do not allow you to draw and fire, sadly because you get a lot of jackasses who don't know what they are doing yet and want to play cowboy. Good news though, you can find training where this is part of the course. You can't just learn where the safety is on a new gun and assume you're good to go. At the minimum, you should find an intensive weekend class of tactical shooting. How will you know if the class was a success? If your thumb is swollen from reloading hundreds of rounds, you have a blister (which is the start of a callous and callouses are good!) on the web of your strong hand, and you have friction rash on your holster side from practicing drawing and pointing (not aiming) over, and over, and over. Also, you should think that while the instructor was a good guy, he was also a stickler and wouldn't let you slide one bit on technique or skill. Think it's overkill? Not cool? Not what SEALs do? Let me refresh your memory from the beginning of this book: "*A lot of people think that the life of a SEAL is non-stop action and excitement and pew-pew-pew'ing. Realistically, 99% of it is train, train, train, then train some more. It is monotonous, boring, and often painful.*"

For long-guns, it is very practical to get some type of sling or strap. You don't need to carry your weapon at the ready at all times, that would be a hinderance usually. Your handgun is there for quick access so you can quickly return fire while on your way to some cover or concealment and get your long-gun or shotgun ready. However, if you're in a bad area and know that enemy contact is possible or probable, a sling that can hold your weapon on your chest so it's at-the-ready could be the difference between life and death. A simple 1-point sling can keep your shotgun or long-run ready to fire in a split second, but leaves your hands free.

Suggestions: Lots of choices in brands honestly and I have quite a few. Frankly, a holster can only have so many technological advances, but one that is lightweight, fit for your specific weapon, and you've *practiced with it* is the best. As a starting point for brands, I personally like Safariland Holsters (Safariland.com) as they have good quality and plethora of options, many of which are modular and can be customized for your needs and fit. Lots of other good gear and accessories too worth checking out. Go to your friendly local gun shop and check out a few different brands and styles to see what you like. It's a very personal thing. Really, I could care less what brand you pick as long as you feel comfortable with it and *have practiced with it.*

Oh yea, have you practiced with it?

One more thing, you need to practice with it.

I highly recommend you take it to the range and practice drawing and firing (with instruction first).

As a final thought, please practice with it.

The good ol' "GO BAG"

Obviously, a go-bag is where you'd put your gear and is somewhat all encapsulating as a concept, but it deserves a mention itself. If you only have only a literal few minutes to get the hell out, you need a go-bag already packed and ready. If you don't have anything ready, your situation is already FUBAR'd (F***ed Up Beyond All Recognition). Also, I suggest you tamper-proof your bag so people (yourself included) don't "borrow" supplies from it and forget to replace it. Just a little string or something to tie the zippers together, not theft proof. Set it up and secure it so it's always 100% ready to go. Check it annually in case you need to replace anything that has expired (food or medicine, etc).

If you look online, you can find some go-bags that are already filled with gear. They aren't bad, and can come at an affordable price of $50-100 usually, but they include some unnecessary things that you can (and should) toss out, and then will need to add some missing items in. Point being, if you do not want to have to collect everything

\\ GEAR

individually from the ground up, it's a good option because anything is better than nothing, but I feel you can get what you pay for.

My reasoning for the pre-made kits not being 100% ideal is that they are designed for modern survival...that is, to get you out of the wilderness and to the hospital. This means you will *absolutely* have to supplement your kit, specifically with consumable products such as medicine (Advil, etc). A little tear-pack of 2 Advil is great for modern survival but isn't much good for years at a time. You want as much of that as you can, both for yourself, and factoring that others won't have it if you are in a good Samaritan mood, or it can become something very valuable to barter with.

Suggestions: I personally have my own bag that I custom outfitted. Consider that you might be carrying this thing for days or even weeks on end and what geography and climate you'll be in. I suggest you go to a good outdoor store and try on a few bags. Aim for something "medium" sized so you can fit what you need but won't slow you down too much. Mine also has a good few external pockets for quick access and easier organization. Further, my medical kit is self-contained and kept on top for fastest access.

Now, let's say you stop for a rest, your go-bag is off (but your weapon is clean and ready, right???), and you get ambushed. You fight your way to safety, but don't have the opportunity to grab your gear. You are now left with what's in your pockets. In this theoretical, if I could only have 3 things on my person, they would be:

1. **Life-Straw.** This allows you to drink directly from the source of any water, no matter how dirty (just not salt-water).
2. **Extra magazine** for my handgun, which is obviously in my holder on my hip.

3. **Quick-clot or bandage** or similar for immediate triage of any wounds.

Also, keep in mind who is in your group. If you have a wife and kid(s), they need to be helping too, even if they can only carry a few pounds, preferably their own food or water.

Fire starters

AT LEAST 3 different methods. I personally have waterproof matches, two different types of lighters, and a magnesium striker. These are cheap and available from almost any Wal-Mart in the outdoors section.

Suggestions: This is easy, really. Any Target or Wal-Mart with a basic "sports & outdoors" section will have this stuff. Don't overthink it.

First-Aid Kits

I have a section devoted to first-aid, but we'll give it a quick mention here in terms of "gear". There are a lot of pre-made kits you can buy, and a lot of them are really great quality…BUT…they are made for short-term use to get you to a hospital. You need to think critically and either build your own kit from the ground-up, or supplement the pre-made kit you buy.

Suggestions: I recommend building your own kit, or investing in a high-end one and storing a couple extra packs or bottles of consumable products elsewhere in your bag (ie – a big bottle of Aspirin, extra quick-clot, etc). Read this book, do your research, THINK about it, make a list, and go to the store and buy what you need. If not, don't go cheap when buying a pre-assembled kit. Mymedic.com has some really good kits, but they are a little pricy and may have stuff you don't need or know how to use.

Tent

Having a tent at all depends on where you are, how many people you are with, how far you are going, and the elements you'll face. If you are on your own, you can do just fine with a small bivouac tent. No frills, but keeps the rain off the one person who can fit in it. If you have a family of 4 or something, you might consider just bringing extra rope and a big lightweight tarp. Sure as hell not as nice as big luxurious tent, but also weighs a lot less and it will keep the elements off you and block the wind. It's a personal decision, because the more you bring, the more you have to carry.

Suggestions: If you need one and are by yourself, get the lightest weight bivuouc type tent you can find. Don't go cheap. If you are in a group or with your family, likewise go for minimal weight and durability, not cheap price. It's not about creature comfort, it's survival, and there is more important gear you want room for, so only get as much tent as you need.

Bug spray

Depends on your climate and region, but if you live in a swamp area, you'll hate life without bug spray. Use it sparingly to make it last.

Suggestions: get something in higher DEET. However, for kids, don't go over 30% DEET concentrate. Picaridin is also a great choice for kids and is safe for babies as young as 2 months. The "Cutter" brand uses Picaridin as the active ingredient. "OFF" brand uses DEET.

Sunscreen

Again, depends on who and where you are, but getting sunburned really sucks. It won't kill you, but makes life miserable and only serves as a distraction.

Suggestions: Obviously, try to avoid sun exposure where you can, including by covering up your skin. I like the little lip-balm style sunscreens for my nose and face. They are lightweight and last a long time, and generally I can cover up the rest of my body with the lightweight clothes I'll be wearing.

Hats

In line with sunscreen, baseball cap is fine, but a big lightweight, broad-brimmed hat like a boonie cover. That can keep the sun off you and remove the need for sunscreen. You might look like a dork, but who cares.

Suggestions: seriously, just a good ol' baseball cap is great. Remember the "tacti-cool" talk we had? If you get a boonie cover, go for something just regular khaki. It doesn't need to be cool, just effective. You'll still look like a dork.

Water

Water might be plentiful, but do you want to drink from that nasty-ass pond water? You need a water purification system beyond boiling water. I highly suggest you get a LifeStraw for each person in your group or family. You can drink straight from any water source and it filters up to 1,000 liters. The average person should drink 2 liters (1/2 gallon) of water per day, so this straw could literally last about a year and half if you used it 100% for your hydration. Lugging around a week's worth of water is a massive amount of extra weight. Obviously, you want a day or two of water at the ready, but any more than that could be over-kill unless you are stuck in the middle of the desert. Buy some purification tabs or system to kill microbes in your water bottle. You can have potable water in a few minutes or hours that way.

GEAR

Suggestions: I would personally have one LifeStraw for each person in your group, plus some iodine tabs. You'll need a sturdy water bottle that can last a long time like a Nalgene bottle. For staying home, you can purchase a heavier duty model from LifeStraw or a similar company for filtering large amounts of water. Clean water might become a good bartering tool for you as not everyone will have thought about that.

Knives

We'll cover this in the weapons section as well, but since a knife will be used more for day-to-day than fights to the death, we will cover it here. First off, you need more than one knife. I suggest a couple mid-range priced folding blades made by a brand like Gerber for around $20-50/each. Just a 3-5 inch blade is good and it will last decently short of severe damage. Keep those as backups for your at least one "really good knife" that should cost you $100 or more, and take great care of that really good knife. Use and abuse the cheaper ones if you absolutely have no alternative.

Suggestions: There are thousands and thousands of options, but for this section I'm making my suggestions for day-to-day use blades. For low-medium level, anything like a Gerber is great. Moving up a little, I like the SOG brand of blades, and Bench Made is regarded as one of the premier knife-makers in the world. I would recommend getting a fixed blade knife that a gut-hook, assuming you intend on hunting your own food from larger game such as deer and hogs.

Food

We'll cover food more, get some dehydrated meals from an outdoors store like REI, Bass Pro, Cabelas, etc. Food takes up more space than you'd think, but I'd recommend 2-3 days of food, even if it's just a meal a day. You might not have time the first few days or even weeks to scavenge or hunt. Hunger can impair your judgment and endurance, so having some calories is important.

Better than dehydrated meals are energy bars, which are universally almost non-perishable (at last for a year or so usually). That said NOT ALL ENERGY BARS ARE CREATED EQUAL. A lot of them are loaded with sugar. You'll get a boost of energy and then crash. That is worse than being hungry. Fueling your body is not a place to say "meh, good enough". Do your freaking research and buy something that has protein and good ingredients. This has nothing to do with fru-fru organic shit, this is survival and keeping your body moving, so be smart.

Suggestions: Personally, I really like Picky Bars. It's good science and good real-food ingredients. Typically, around 200 calories per bar, it has a 4:1 ratio of carbs to protein for good "immediate" energy and then some protein to last. The carbs are longer-acting, meaning not processed sugars that you metabolize and waste. They are only a couple bucks each, and they don't taste like shit either.

Toilet paper

It seems like a luxury item and doesn't seem very SEAL-like to worry about TP, but until you try wiping with pine cones, you won't appreciate how valuable this precious commodity is. Use it wisely and try to stretch it. I'm not pointing fingers, but from experience of living with women, you might need to ration it....Just saying.

Suggestions: buy the thin stuff and try to make it last.

\\ GEAR

Flashlights

Kind of like with knives, I prefer to have 4 or so backups of small, lightweight (preferably aluminum), flashlights that have an LED light and only need a couple triple-A batteries. They'll last forever, weigh very little, and you've got plenty of backups. Do NOT mix them all up though and use all 4 at once, that's a great way to lose track of your gear. Have one flashlight and know where it is at all times, which should be in your hand or in your pocket. I also really like the head-band type lights, they can be really useful and worth carrying in my opinion.

If you don't want to spend the money on more than one, get at least a single good flashlight that has a light-bar or something that can spread non-directional light (like a lamp or a light bulb). This is really useful, but exercise caution as it can give you away. Last tidbit, get a light that has a red filter. They are much harder to see from a long distance and can help keep you hidden.

Suggestions: Way, way too many options to list. Poke around on Amazon or hit up your local Walmart or outdoors store and see what you like. Don't go super-cheap and buy the little 2.99 flashlights by the checkout counter though, you get what you pay for. Aim for something around $10-20/each.

Foil-blankets/mylar blanket/survival blanket

This can be a literal life saver. Very small, very light, and traps heat like you're a thanksgiving turkey. Same thing they hand out to marathon runners after a race so they don't drop body temperature too fast. Get at least 1 for each person in your group.

Suggestions: really doesn't matter in my opinion because practically speaking it's just a giant piece of foil. Get a middle-of-the-road priced option and you'll be set.

Rain gear

This might also seem like a luxury item until it's raining, you have to keep moving, and it's 40 degrees outside. You will die of hypothermia. I suggest getting a NON-thermal insulated rain gear (or poncho, suit, or jacket...whatever you decide on) because that insulation might be more than you need if it's only 60 degrees outside. The exception of course is if you decide to get a rain jacket that is also your regular jacket, in which case that's a great idea.

Suggestions: There are a lot of good brands, but I would suggest something like Frog Toggs. Unlike cheap stuff that is made of plastic and can tear easily, and once torn is impossible to fix, it is made up of a multi-layer material that is designed to resist damage and if punctured, will often resist tearing. Frog Toggs basic model is the "Ultra-Lite2" rain suit and is only $25.

Teeth

You will want a toothbrush and toothpaste. The toothpaste won't last forever, so you might want to stretch it out and only brush with toothpaste once a day or even once a week, depending on how dire of a situation you find yourself in. Do what you can to keep your mouth clean in-between, such as just brushing with no toothpaste. Keep in mind the bristles don't last forever either though.

Suggestion: I don't care what brand you buy, but folks don't recognize how important oral hygiene until you are your own dentist. You'd be remiss to not take seriously "don't forget to floss" too.

Technology & Communication

LOL. What technology? Your iPhone is a paperweight now. Everybody takes communication for granted right up until it's gone.

GEAR

We've become accustomed to having the world at our fingertips, getting pissed when the "service around here sucks" because it only gets 3G. Assuming you can even keep charging your phone, the infrastructure to make it work will be shut down or even physically destroyed. Frankly, I would use my phone for the pictures on it: go through occasionally until your battery runs out. It could help you keep motivated on why you are fighting and surviving, if you have something important to get back to, like your family.

Technology substitutes:
Maps

To the point though of "technology", you need to get back to survival mentality of 1950's tech. For example, do you have a Rand McNally map? Get one. I get a new one every couple of years. Crazy right? A real map printing on paper. If you don't know how to use it, grab a beer and sit down with your wife and kids, whoever, and ask 'Hey, this is what a map looks like for one. Tell me where we're at?" It sounds simple, but you need to have it and a lot of people will overlook it and regret it sorely. The reason is that the highway system will stay the same, all the interstates, back and forth, East and West. You can use that to plan routes and avoid cities that need to be avoided.

Suggestions: get both a United States map and one of your state. Try to get one that is weather-proofed and durable.

Compass

It's good to know that the sun rises in the East and sets in the West, and knowing which one is the North Star is great, but survival experts know that it takes a lot more than that to even halfway navigate. Depending on where you are in the world and what time of year, the sun doesn't come up exactly in the East, nor set exactly in the West. Likewise, what about when it's noon?

Suggestions: Even just a tiny compass on a bracelet is better than nothing, but if you really want to be able to do land-navigation you'll need a good sighting compass. You'll also need to learn how to use it which is more training than I'll be able to cover in this book. You're training to be self-reliant, right? Surely you can Google "how to use sighting compass and map".

AM/FM Radio

Now, some more basic 'technology' will survive, although hardly anyone uses it anymore. A good old AM/FM radio with a hand-crank or solar panel is a great thing to have. The AM/FM radio broadcast is far less complicated and while it might go down temporarily, will survive for quite a while longer overall, and survivor groups will inevitably start running them at some point. If you're walking from the West Coast to get to family in the Midwest, and Denver's gone down the shitter, you'd want to know to avoid Denver. Information will become invaluable. Knowledge of any kind can help you survive longer.

Suggestions: There aren't any specific brands I suggest, but don't go too cheap and try to look at some reviews. Bonus, a lot of these "survival" radios come with a built-in flashlight.

No-Go technology

Some of you might be gravitating back towards better technology like "what about satellite phones? The satellites will still be there, right?!" First question: who the hell are you going to call? Secondly, it seems relatively simple that a signal bounces off the satellite and bounces back down. Let me assure you, it's far more complicated than that. The

\\ GEAR

hard truth is you must accept that communications as we know it today would be the first thing to go, so you might as well plan for that, not spend time trying to fight it. Any kind of electronics that runs on batteries even, all that kind of stuff will go, incrementally. It will just go away. Plus, if you are on the move, the longer you think you have to have it, you're going to carry your battery or your generator with you.

Miscellaneous, but important stuff:

There are some things that don't immediately come to mind when you think about "survival", but especially if/when you are able to make a permanent defensible location, there are some things you'll want to have or find a way to obtain. For example, you want to get a fire extinguisher. If you don't have one, there's a good chance other people will overlook them in looted stores.

Think creatively here for things you'd want or need that you can scavenge. Another example of a completely looted store with no merchandise left. What about all the racks of shelving? That could be extremely useful for a variety of reasons. It's typically structural grade stuff and can hold a lot of weight, meaning it could be used for blocking off windows or other defenses.

The point here is you have to think outside the box. I cannot think of everything YOU might possible need. I don't know you, your situation, your location, your skills, your weaknesses, or anything else. If you're going to survive, you have to be SELF RELIANT which by definition means you can't pick up a book to get the answer in the heat of the moment. This book will help you get in the mind-set to figure out the answers, but won't answer the questions for you.

TLDR:

Don't wear or do things to stand out, you want to blend in. Consider your climate for clothes you need and those to leave. Always KISS (keep it simple stupid) when selecting gear options. Don't get anything requiring technology when possible. Most importantly, don't forget or skip over the basics. Think: beans, band aids, and bullets.

Firearms

Sticks and stones may break my bones but hollow points expand on impact.

I bet half of you skip right to this section. It's OK, I don't blame you. No big introduction needed here, let's jump into it. I'm going to go over the main categories of firearms generally available and list out on a 1-10 scale (10 being best) of the various factors of each category. Feel free to disagree with me here if you're a gunner, these are just ballpark references for the rest of us.

Semi-Auto Pistols – Pew Pew Pew

- Round capacity: 9/10
- Maximum effective range: 4/10
- Weight/portability: 10/10
- Close-quarter combat effectiveness: 9/10
- Accuracy: 7/10
- Reliability: 8/10

Firearms

Suggestions: No shock here for anyone that knows anything about firearms, Glock is a great choice. They're pretty much everywhere. Sig Sauer has some great handguns, most team guys love Sig's because that's what we use in the Navy. Really though, it's about practicality, and in this case guns use bullets and bullets are a limited commodity. When you're out of bullets, you have a paperweight. Because of that, any pistol that's chambered in 9mm is a good choice. Why? 9mm is practically a universal round, a lot of people carry it, so the chances of you being able to find more ammo (or take it from someone) is really good. But if during your travels you stumble upon a .45 Kimber, and a bunch of .45 rounds, take them anyways.

Now, this is not to say that only Glock is good, and only 9mm is good. To those who ask "9mm or .45?" I say "yes". The two most important questions about the side-arm you choose is:

1. Are you proficient with it?
2. Do you trust it with your life?

If you aren't proficient with it, you can't know how trustworthy it is. Does it get jammed? Does it have a lot of small parts that have to be cleaned often? A rule of thumb is you should put AT LEAST a few hundred rounds through a weapon before you trust it. Avoid cheapo brands as well as any weapon chambered in a less common round such as a .357 Sig because it will be harder to find replacement ammo. Sorry, no Desert Eagle .50AE's either. They are seriously bad-ass, but it is a hand-cannon better suited for demonic grizzly bear zombies, not regular people zombies.

Revolvers – The classic wheel gun, partner.

- Round capacity: 5/10
- Maximum effective range: 5/10

- Weight/portability: 7/10
- Close-quarter combat effectiveness: 8/10
- Accuracy: 8/10
- Reliability: 10/10

I love the simplicity of revolvers. Jams are impossible, they are damn near indestructible, accuracy is very solid depending on length of barrel, and ammunition is plentiful as long as you stick with something like a .357, .38 spc, .45 colt, etc. Some are chambered to fire 9mm, which is a bonus to be interchangeable with a semi-auto pistol. Overall, it's a classic example of: "if it ain't broke, don't fix it". Revolvers have been around since 1835 and there is a reason why they still exist with only minimal mechanical improvements. Pull the bang-switch and it goes boom, every time.

The downside is that they are a bit heavier than a pistol, especially semi's with a polymer frame, and they hold fewer rounds. A 9mm semi-auto can handle anywhere from around 15-20 rounds depending on the make and model, whereas the best you can get with a revolver is maybe 8 rounds. While shot-for-shot, they are equally effective at close range, revolvers lose a point in close quarters combat for the decreased round capacity. Keep in mind though that being out in the wild with no replacement parts or possibly even tools to maintain a weapon, and the 10/10 reliability makes this a solid option.

Suggestions: Lots of good choices honestly, brand is not nearly as important with revolvers. In fact, it's damn near impossible to list out all the various models made by all the manufacturers. However, if you stick with a manufacturer like Colt, Ruger, or Smith & Wesson, there is a strong chance you'll have a fine weapon. Another bonus is that revolvers are a safe bet to buy used.

\\ Firearms

Stay away from: overly big rounds. You don't need a S&W 500. It is a hand cannon and totally unnecessary for pure survival (although it is fun as hell to shoot). They are literally made for defense against Grizzly bears.

Shotguns – Big, Ugly, and Dumb. Like your mom (ha!)

- Round capacity: 6/10
- Maximum effective range: 7/10
- Weight/portability: 7/10
- Close-quarter combat effectiveness: 10/10
- Accuracy: 10/10
- Reliability: 7/10

We have to distinguish between a few sub-options however. The break-action, pump-action, and autoloader.
A break-action is the classic double-barrel typically used by hunters. It's very simple mechanically, it's quick to reload, but comes at the cost of only holding 2 shells at a time.

PRO TIP: There is one important caveat here, and that's that you must be familiar with the break-action shot gun you have, and the ammo you use in it. It's not uncommon for the metal base of cheaper shells to expand from the heat and become difficult to dislodge in break-actions. I have a Ruger Red Label that is a fantastic hunting gun, but certain shells tend to get stuck after it warms up from a few shots. Just about any other round, I could damn near melt the barrel before any issues come up.

Auto-loaders are essentially a semi-automatic shotgun. You pull the trigger and you will both fire a shell and load the next shell, no pumping required. Typically a good capacity of at least 3 in the tube and 1 in the chamber, but personal defense weapons can hold 7 (or more in some unique cases).

The downside is there are more moving parts, so it's more susceptible to jamming or having mechanical issues. This really depends on what brand you buy and taking care of it. Reloading isn't horrible, but it is one at a time. You'd want to practice running out of shells in a firefight and having time to reload only 1 or 2 at a time before needing to swing up and fire again.

The pump-action has been around for a long-time and is probably the most common option. It blends the best of all worlds in that the mechanical operation is more simple to fix, it holds just as many shells as the autoloader, and it's damn-near as fast firing rate if you practice. In fairness though, ANYTHING with moving parts can be a liability, but I feel that it's a small mark in the "con" column here. The 'crunch-crunch' of racking a pump action is a great sound and some argue it's an "intimidation factor", but I wouldn't count on that and keep in mind that the sound is a liability if you're trying to be quiet.

Suggestions: All three choices work, and what matters most is what YOU are comfortable with in terms of your own specific firearm. However, I will tell you I 100% go with the pump-action. More specifically, I would suggest the Remington 870. It has been around for decades and is trusted by nearly every law enforcement agency and military branch out there. It comes in dozens of variations, and there are numerous accessories and kits to customize it to your needs. Best of all, it's quite inexpensive considering its extraordinary reliability, and since it's so common (ie – there have been over 11,000,000 sold since it was first developed in the 1950's), finding parts is no problem.

Stay away from cheap, off-brand, "tactical-style" looking shotguns you find at Wal-Mart for a couple hundred bucks with no warranty. All show, no go. Likewise, while I love the concept and it definitely has the 'cool-factor", I'm not 100% sold on some of the new options such as the Kel-Tec KSG or IWI Tavor. For those unfamiliar, the KSG

\\ Firearms

shotgun shows up in the movie John Wick and is a double-magazine pump action, meaning the gun draws from two tubes each holding 6 shells, for a total of 12 shells. Put in stubby shells, and it can be upwards of 20 shell capacity. That's impressive, but this is a pretty new concept. I've seen and heard good and even great reviews, but like I said, it's a personal choice to decide what weapon to trust your life with. Contrast that with a choice like the Remington 870. BEST GUN EVER in my opinion. If you want though, get one, but take it to the range and punish it to see if it meets your standards. Just pump box after box after box through it. Last thing though, those KSG's are so short you have to be really, really careful to not let your hand slip out in front of the barrel. Pumping a shotgun is much easier with both hands intact.

Lastly, stick with 12 gauge, or 20 gauge if you really prefer. All the other stuff is far less common and, in some cases, far less effective. I would guess that the 20ga is about 60% as common as a 12ga, and I appreciate the 20ga because the trade-off of less "knock-down power" in favor of lighter weight and slightly smaller shells so you can carry more shots. The compromise in ballistic effectiveness is truly minimal. This is important too depending on who is shooting it; a smaller framed person would benefit from a 20ga. If you are on the fence, definitely go with 12ga though.

Long Guns - when you need to reach out and touch someone

- Round capacity: 10/10
- Maximum effective range: 10/10
- Weight/portability: 6/10
- Close-quarter combat effectiveness: 3/10
- Accuracy: 10/10
- Reliability: 8/10

This is a tough one to weight scores because there are a lot of options for rifles. Some only have capacity for a few rounds, but are extremely accurate to very long distances. Others can accommodate magazines with 30+ rounds, and accuracy is dependent on barrel length. Overall though, the rifle is untouchable for maximum effective distance, is generally extremely reliable (especially a bolt-action), and ammunition is plentiful if you choose a common hunting round. The downside is a good long-distance rifle is worthless in close combat, it can be pretty heavy, and they are very loud when fired.

Suggestions: 30.06 bolt action. Grandpas old deer rifle is just fine.

<CUE THE AUTISTIC SCREECHING> OK, OK, calm down now. Let's talk about the AR. An AR-15 platform weapon is just dandy, I have shot one once or twice during my career, so suffice it to say I know what the hell I'm talking about. Here's why I'd choose something else: First, I'm assuming I'll have a handgun and a shotgun in this hypothetical, so I need something specifically for accuracy at longer distances. An AR is a weapon that is born of compromise and designed for a specific purpose of warfighting:

- It has to be issued to hundreds of thousands of infantrymen for God only knows what missions or climates or geography or enemy.

- It needs to be light-enough that it can be carried by hand, but sturdy enough that it doesn't break

- it needs a round that can be effective to a distance for infantry combat, which generally is only a few hundred feet on average.

- It needs a barrel length that allows for clearing rooms and relatively close-quarter combat, but just long enough that accuracy at the top of the effective max range isn't totally worthless.

In fairness, you could argue all of those items are GOOD compromises for survival. Like I said, it's a great weapon, they are a hell of a lot of fun, but it's a compromise because in every category, there is another weapon better suited (long distance, close-quarter combat, etc) but the AR does the job for the purpose of warfighting. A long-gun for long-gun's sake, like a good bolt-action with a nice scope makes zero compromises for what it's intended purpose is. So, if you've got a pistol and a shotgun, don't compromise here just because you love your AR.

Summary: The big question: if you could only choose one gun, what would it be? Easy answer for me (drumroll please……): Shotgun

Across the board, it is suitable for anything a rifle or pistol can do without compromising too much. As long as you don't have a long-barreled goose gun in 10ga, you get:

- versatility of rounds.
- A slug will take down a deer at quite a few yards distance, no problem.
- Use bird shot and you can kill a duck or wild bird for dinner.
- Heavy buck shot for self-defense/daily use.
- Overpenetration is a limited concern, unlike a .223/5.56NATO used in the AR. For those not following, you don't worry about the bullet going through the bad guy, through a wall, and into your family member.

- Capacity, assuming you have my recommended pump-action, could be as many as 7 which isn't horrible. I feel it's offset by being able to quickly reload a single shell if needed, as opposed to removing a magazine, reloading the magazine, and reinserting (assuming you've already emptied your back-up magazines).

Now, for you AR-lovers, my 2nd choice would be my go-to AR that some might call their "truck gun" that can get banged up and still go "boom". Another weapon I really like is the classic Ruger Ranch Rifle. It's a seriously rugged little rifle and Ruger did a great job in the design.

Another big caveat is the AR-10 which shoots a .308 or 7.62x51 (effectively the same bullet for purposes here) which is also a NATO round meaning it should be easier to find. The .308 is the cartridge most every other hunting cartridge is compared to. The AR-10 looks basically identical to an AR-15, it just shoots a bigger bullet very similar to the 30.06 bolt action I mentioned. Only thing is they are much less common, but if you've got one, good for you.

The shotgun wins for me though because I put close-quarter combat as a higher priority than a 300 yard shot capability, plus a chest full of .00 Buck is a hell of a way to put bad guys down fast. Picking a gun is a very personal preference thing, like drinking wine. You might like one bottle of wine, but I might not like it because my palate's different. The two most important factors:

1. Accuracy and shot placement are far more important than what round you're firing.
2. The weapon you are confident and proficient with is far more important than anything I say.

\\ Firearms

TLDR:

- Pistol/Revolver: Semi-auto (Glock or similar) in 9mm with a 4 or 5" barrel.
- Shotgun: pump-action with at least 5 round capacity in the tube. .00 Buck shot for self-defense, and have a couple slugs for hunting deer, and bird shot for ducks or wild bird.
- Long gun: grandpas 30.06 bolt-action. Remington 700 is a great weapon.
- If I can only have one gun?
  - First choice: Shotgun
  - Runner up: AR-15 variant

PRO TIP: What is proficient? If a semi-auto handgun, can you disassemble and reassemble blind-folded? Do you go to the range at least once every couple months? If yes to both, you are proficient. Disassembling and reassembling a handgun while blindfolded might seem overkill, but you should be that familiar with how it feels to strip, clean, and put a weapon back together. You don't want to be fumbling around trying to figure out what the doo-hicky doesn't fit in the thing-ama-jigger.

When you run out of bullets

"Conan, what is best in life?"

"To crush your enemies, see them driven before you, and to hear the lamentations of their women"

-CONAN THE BARBARIAN

Hopefully you've got plenty of ammo and your weapon stays functional, but you should have already been in the mindset of "hope for the best but plan for the worst". So, what do you do when your firearm turns into a paperweight? Time to get medieval.

Knives

We covered knives in the "Gear" section for day to day use. There are a million options out there, but when it comes to self defense, I'd prefer a no-frills, but well-made sheath knife like a KA-Bar. Alternatively, there are some fantastic folding knives out there with some pretty good-sized blades on them. Those are nice because they can fit in your pocket still for quick access and pop open fast. Be wary of "cool" looking knives, usually a lot of show and no go. I personally stay away from heavy serrations because it gets caught up on rib-bones, which is a hassle when fighting off multiple enemies.

The amount of fighting you will do with your knife is limited. I don't suggest hand-to-hand combat with zombies, but it is nice to be prepared with the best options. If you come across enemies of the human variety and you're fresh out of guns, you will literally be bringing a knife to a gun fight so it's going to have to be a good one. The primary uses of your knife will be cutting stuff whether it's heavy brush, skinning fish, or your hair. What, did you think there is going to be a barber walking around charging two bits for a shave and a haircut? A little tip I unfortunately learned from my dad…. Be sure to sterilize your knife after you use it to trim your toenails, before you use it to chop vegetables. I know, Apocalypse truths can be painful.

A general rule of the zombie apocalypse: Never trust a man with your life, your wife, or your knife. You keep that thing holstered or in your hands only. Someone wants to use it to skin a deer? No worries, you have time. They need to cut a tarp to make some shade? You're happy to help. Some pretty thing bats her eyes and wants to borrow your knife to trim her bangs? Nope. That will turn into a weekly chore. Sorry, Mary, you're going to have to grow that hot mess out.

Suggestions: SOG sogknives.com are well constructed with good heavy blades and come extremely well-sharpened from the factory. I have a KIKU model I like. Of course, Bench Made is another great choice, but can be a little higher priced. Even more than guns, a knife is a very personal choice. I recommend you go to a store and actually hold the knife, feel the weight, see how easy it is to open/close or clear the leather of the sheath. Once you choose one, carry it around and use it so you get familiar with it.

Thinking outside the box

Let's think outside the box, beyond the obvious bats and knives. It's good to be creative as long as it's practical, but it is most important to analyze a potential weapon's strengths and weaknesses, just like we did with firearms.

- How heavy is it?
- Can you carry it long distances?
- Are you walking, in a car, holed up?
- Can you conceal it?
- Can you swing it multiple times without getting too worn out?
- It is consumable or reliant on something consumable that could run out?
- Does it need to be sharp? Can you re-sharpen it?
- How close in do you have to be for the weapon to be effective.

Now, what if I need to fashion a weapon? Let's use a morbid scenario-game: You and other people are locked in Home Depot, it's a fight to the death, only one person can leave. You get to pick two items in the store. What do you pick? It's a pretty demented "game" to pretend, but there are different strategies and once you start thinking about it in depth, you realize it is a good way to play devil's advocate against yourself.

People have all kinds of crazy different ideas, like, 'I'm going to get a chainsaw and premixed gas. That's two items and is definitely lethal.' Other people would counter "That's dumb. It's heavy, it's got a limited supply of fuel, and I'll hear you all the way across the store. I'm going to get a lighter and a propane tank, hunker down behind some good cover, and blow everything up as soon as you get close to me." There are pros and cons to consider in everything when it comes to survival and combat. There's always a give-and-take. Self-defense and weapons are no exception. In essence, anything you can pick up could be used as a weapon. If you need something that instant, a rock right by your foot is better than a knife or gun 100 yards away. Go back to those baseline questions: is it heavy, or light? An axe is sure deadly and scary, but it's going to wear you out swinging it and you don't have a lot of speed or control.

I'd stick with something simple. Probably head to the garden implement section and grab a decent machete or even just a sturdy hickory stick. Hickory is an extremely hard wood that lasts forever and could be sharpened on one end as well. It's not too heavy, it doesn't have parts that break, and it works. Not cool or glamourous, but extremely functional and practical.

Another form of self-defense to consider is protection. I had a buddy once say he would hide as long as possible and choose his weapons at the last minute. After we mocked him endlessly for being such a wimp, I had to admit that's a smart idea for the purely chaotic, hypothetical situation I had proposed. Just like in so many movies when the heroes are being chased by multiple threats – let's say the mob and the cops. If they can create a stand-off where the mob and cops are fighting each other, enemy is killing enemy and our heroes can escape. It's a win-win-win situation for the good guys!

Same thing with the zombie apocalypse or the Home Depot death trap game. If you can lead people trying to kill you to another group of people (or monsters) trying to kill you and convince them to kill each other instead… fantastic. Why wouldn't you camp out in a safe spot and enjoy the carnage that's making your life easier.

For those who watch "The Walking Dead" on AMC, you know that the bad guy Neegan favors a Louisville Slugger wrapped in barbed wire he affectionately named "Lucile". Not necessarily a bad choice. It sure as hell has the intimidation factor going for it, but personally, I think that a longer hickory stick is a better choice. Just like in boxing, there is a reason a guy with fractionally longer arm reach has a big advantage in that fight. Now think of having a 3-foot extra reach over a baseball bat. That's a factor worth considering. I also don't love the idea of barbed wire. When the main target is a zombie skull covered with rotting flesh, I'm avoiding anything that may get stuck in there and need be pried out. Just gross.

Here's the most important lesson though: If you don't have a gun, your goal with whatever weapon you have is to get to a gun. Use that gun to get to a better gun and/or more ammo. Whoever has a gun has huge advantage over you, but since you can't control that, use what you have to your advantage which could be stealth or cunning. Remember: Improvise, adapt, and overcome.

If you come across grenades, use at your own risk. Sure, they're relatively lightweight, transportable, and can inflict a lot of damage… on WHOEVER is around the vicinity detonation (yes, I most definitely think that will be you and everyone you care about). My favorite part about bats or sticks? Everyone can swing them safely. If someone hurts themselves swinging a baseball bat, sorry, Darwin, they ain't fit for survival. Knives are also for everyone. Guns simply have to be for everyone, they're too important to not know how to properly use, store, and clean. Just because you may have access to things like grenades, missiles, tanks, IEDs, or bulldozers doesn't mean you should use them. In fact, this SEAL highly recommends that you don't. When it comes to any type of apocalypse follow the simple KISS guideline.

TLDR:

Don't choose a fighting knife just because it looks "cool". If you have to fashion your own weapon, be practical. A chain-saw is scary, but it's loud and heavy and you have to have fuel to make it work. A simple

\\ When you run out of bullets

6-foot long hickory stick is better. For those unfamiliar, Hickory is an extremely hard and durable wood. You do NOT want to get popped in the side of the one with one. Don't believe me?

TACTICS and TIPS

"Those who live by the sword are shot at by those who don't".

Let's talk about defensible locations and what are some tactics you should employ to defend yourself. Probably most importantly is the ability to tip the scales to your favor before the bullets start flying. If you have the high-ground, cover or concealment, and advance notice of the enemy's movements, your odds of survival and victory are significantly increased.

Let's use a silly example. You see a small uninhabited island in the middle of a lake. Seems great huh? Unlike a guy who owns Google or Amazon and your island is already populated with all the food and water for 20 years, and its miles and miles off the coast, an island is probably a dumb idea. Sure, it's "defendable" because it's surrounded by water, but it could be a prison too. Likewise, if it's big enough to have food and a fresh water source, you'll have your work cut out defending it and keeping it "yours". So if you haven't thought about how to grow or hunt your own food, how you'll know if someone intrudes, how to get fresh water, you'll just starve to death or die. You've traded one problem for a much bigger problem.

A more realistic situation is your own house. You have to weigh the factors of how long you can stay there and how long the situation will last. Do you think it will be contained and over in 30 days, or are we back to the stone-ages? Those are the first questions to answer, which will help you decide if and how long you should plan on staying, and then what you need to do.

If you're waiting on me to give you the answer, again I tell you, you have to be self-reliant. Every single house and every single situation is different. I cannot give you universal answers, however, I can give some broad stroke thoughts and suggestions.

First, if roads are still drivable, I'd have my vehicle backed in somehow so I can hop in and peel out if needed. Depending on if you have a garage, you might have to back it up to your front door. This is where cover and concealment comes in handy. Get some tarps or a bunch of shit and cover up your vehicle. People will know there is a car there, but they won't see through what you're concealing, so it could give you a precious few seconds needed to hop in without getting shot.

Next, do not make the mistake of thinking that your house is safe just because you board it up. Why? You are in a literal tinder-box and can get smoked out or burned alive. It's good to keep the bad guys out, but can YOU get out if you need to? Furthermore, can you get out without walking into an ambush?

On that note, you always want to have two escape routes, and not on the same side of your house. Ideally, if you have a hidden escape door, that would be great. This isn't James Bond kind of stuff, I'm talking about just covering up a hole in the wall with a chunk of plywood and branches, that you could crawl through. No one will assume to look for it, they'll be covering doors and maybe windows.

Your Team

Here's how it works in the teams: everyone is proficient in everything, but everyone is also an expert in something specific. For example, one guy knows every weapon known to man, how to shoot it and how to fix it. Another guy has advanced life-saving and medical training, but everyone has been through some pretty extensive training too.

That's how you need to train your team, and there's a good chance you'll have to do the training.

One thing for sure though, *everyone* should know how to shoot *safely*. Period. That is your first task, to make sure everyone knows all the weapons in your group, so they can pick it up and start putting lead down-range. Ideally, they'd be able to break it down, clean it, and reassemble quickly with no assistance…but at least point and shoot it.

Definitely want a medical person if at all possible, but at least knowing CPR and minor first-aid is better than nothing. It's your responsibility right now to find and secure some training. It has nothing to do with getting a certificate or a card you get to put in your wallet, you need to memorize and engrain that training into your brain like your life depends on it. **I cannot emphasize this enough**…do not skip the basics! In an apocalypse, more people will die from basic things like cardiac arrest, infection, dehydration, etc. than zombie bites. In fact, right now, right here, take a break and go here and sign up. I'm not kidding, put the book down RIGHT NOW DAMNIT! Go here: www.redcross.org/take-a-class/cpr/performing-cpr/cpr-steps.

To make it even easier, here's an abbreviated break-down on how to perform CPR:

TACTICS and TIPS

1. 30 chest pumps.
   a. STRONG ones, you aren't tickling them. Push down at least 2 inches.
   b. Go to the beat of Bee Gee's song, Staying Alive ("I, I, I, I'm staying alive, staying alive).

2. 2 breaths.
   a. Tilt their head back slightly, pinch the nose, and breath for at least 1 second to see their chest rise.

3. Repeat steps 1 and 2 for as long as you can.
   a. Most people do not realize that performing CPR is extremely exhausting. If you can continue for more than a few minutes, you are stud. Ideally, the victim is revived before then, or you have someone to swap with.

If you don't have the luxury of handpicking your team, then you want to teach the skills you do know and try to give people jobs. Keep them active and thinking proactively. Not fixating on the bad stuff and getting paralyzed with fear. There's a good chance you'll be moving towards somewhere you need to get to or hunkered down with nothing to do. NEVER waste downtime. Ever. There is an old moto poster from the Marine Corps that says "killing time kills Marines", same applies here.

Just start talking about things. 'Hey, what happens if I break my leg? What are you going to do?' And make that person think it through and recall what you've taught them. Whoever has skills, they take a turn teaching what they can. We're walking, we're teaching, we're bullshitting, it takes your mind off of the situation, you're just learning something, but you're still moving towards the objective.

As far as who does what, there's no wrong thing to do here. Obviously, use the skills people have to everyone's benefit, but also beat into your team that everyone needs to be "on" unless it is their turn to sleep. In the teams, it was just understood: if we get hit right now, you have a rally point within 30 seconds you're going to go to. And that changes the next 30 seconds. It's a constant in your mind, you're always thinking. Yes, it can be mentally draining, but it also keeps you alive. It's not as easy as it might seem because thanks to technology, you could walk up to most people today and stab them in the neck and they wouldn't know until it happened because their nose is stuck an inch from their iPhone. You'll have to train that out of both yourself and others.

Clearing rooms

The thing about clearing rooms is you are in a big disadvantage, especially if the person on the inside knows you're coming. This takes a lot of practice and training with a well-equipped team of equally well-trained shooters. Frankly, if you haven't been trained and practiced extensively, you'll suck at it. Only do this if you absolutely have to (ie – you'll starve otherwise or something). If you must do it though, the most important thing is to act with overwhelmingly violence and speed. This isn't a peek-a-boo. Here's the basics:

If you're by yourself, and the door is open (hopefully), stay a few feet away from the wall and make an arc around the open doorway to visually clear as much as the room as possible without entering. The two corners along the walls to you that you cannot see will be blind to you, so next, if the door is halfway open and thus providing cover to one side, quickly poke your head in with your weapon leading the direction of that corner. Lastly, do the same thing around the open door.

Keep in mind that unless you have ninja skills, you will make SOME noise, and a person could easily shoot through that door because it is COVER, not concealment, remember? To mitigate odds of catching

TACTICS and TIPS

a bullet, you could try being sneaky and get real low to the ground for that first peek around the door jamb.

Now what if the door is closed? Try to think like if you were the one in the room if you knew an ambush was coming. You'd probably hide in the corner right? Well, if you have to violently breach a door and you're on your own, clear those most dangerous corners first and as fast as possible. Try to ignore the impulse to look at anything else. With any luck though, you'll catch the bad-guys off guard and those precious few seconds can make all the difference in the world.

If you have a team, simply divide the room up to 4 sectors. This takes real discipline, because the first person in will button-hook around the door jamb to secure the corners where the bad-guys would most likely hide, like we just talked about. That takes discipline and trust because those first two people have to ignore and not engage anything else in the room, trusting that in the split second later, the next guys in will cover them and shoot any threats in the far back corners.

Like I said, clearing rooms suck, especially if you don't have extensive training and practice. Super especially if you don't have extensive training and practice with the specific team you are working with. Super-duper-extra sucky if you are by yourself.

Only do this if you really, really have to.

TLDR:

Like always, stick with the basics. Don't make overly complicated plans, and don't make plans without contingencies. When you've got downtime and aren't sleeping or resting, keep your team busy with meaningful jobs or talking about proactive theoretical situations. If everyone is engaged and bought into the plan, the team will mesh, minds will meld, and tactics will improve.

FOOD

"Never eat more than you can lift"

– MISS PIGGY

I personally have a years' supply for two people (I know, I'm such a romantic) in the basement that I bought from one of the Utah prepper type websites. They've done all the research to make it last, and it will stay good for literal decades. It comes in little totes, I have around a dozen of them, and you can carry it and move it around. It may not be the best food, but you'll survive, and all you need is water to rehydrate it. I can also tell you, it is waaaay better than an MRE, which if you've never eaten, you're lucky. We'll cover water in the next chapter.

Keep in mind that how much you can carry is directly proportional to what you eat. If you've got an 80 lb pack on your back, it better be full of plenty of calorie-rich food because you are going to need to fuel up constantly. If it's full of heads of lettuce, you might be up you-know-what creek without a paddle (but then again, "that guy" would probably have a paddle in that stupid heavy bag. Don't be 'that guy').

\\ FOOD

Before I break into the dehydrated stuff, like most Americans I have a pantry with canned and dry goods. I make sure to always have a pretty good stock of rice and pasta, plenty of beans, and vegetables. I rotate through them as we use them up on a normal-life basis so that the newest cans are in the back, oldest move up from to be used first. It would not be what I call gourmet, but we could have plenty of food for at least a week or two just on the pantry shelves. As an absolute minimum, anybody should have a minimum week's supply, just for earthquakes or natural disasters or power outages.

Despite all the chaos that is going on the world, most Americans feel unusually safe and trusting of the Government, science, nature, and each other. Not me. In the likely event of a minor disaster, I'm going to be comfortable. In the not as unlikely as you think event of a major disaster, I'm going to survive without having to exert myself too much.

Okay, I admit, a year is long-time to stock up for, but what if the event lasts longer than a year and the supplies run out? You need to have made that kind of decision and plan way earlier than a year, you can't wait until you're already hungry. I wrote this book though for the mid to long-term survival. I'm not getting too far into how to build a mud hut and gather berries, because there are a million books on that already. That said, unless you're a master hunter, fisherman, gardening, or a hippie that was raised on foraging, obtaining food won't be as easy as you think so while you figure that out, you better have some rations on hand.

Be prepared. Get fed. Stay alive. Some things are just that simple.

Growing your Food

You have all that freeze-dried food to keep you going while you get your crops growing. Even in good climate with optimal soil, any vegetable or fruit will take at least 60 days to produce edible results.

That depends on what time of year it is and where you are though. If you try planting spinach in Michigan during January, then maybe you should go out and hunt while someone else does the farming. It's not for you.

Harkening back to earlier chapters, you need to decide if you are staying or going first. It would be a horrible waste to start planting all your seeds, then have to bail. Before you grow anything, you need to make sure you can defend it and plan on staying a long-time.

You'll need to do your research to determine what plants grow best in your region, but I would stick with good root vegetables and squashes as a baseline since they tend to be more "dummy-proof" and are not as susceptible to wild animals and birds eating them. Be smart about your choices, you may not need the variety you think you do. Do you plan on making a meal of jalapenos? If not, skip those peppers and plant more carrots or potatoes. We're going for calories and survival here, not spice and flavor.

You can buy "survival" seed packs, but I find that those can often have more stuff than you really need, or things that won't grow in your region or soil type. Do some research before you buy a pack of 50 different plants, because realistically, are you actually going to grow that many different things? Do you even have the space to do it?

Let's not forget that even in an apocalypse, life is for the living. If you've got kids, this is a great family activity to teach and have fun outside. Little guys will love it when they see the first green sprouts popping out of the dirt and knowing they helped grow it. That will help them eat veggies too (not like they'll have a whole lot of choice). A good mud-ball fight is always good times…you are allowed to have some fun, even in a zombie apocalypse.

Basic crop suggestions (although you can add on whatever you want):

- **Corn.** It has lots of calories, is pretty darn sturdy, and grows without much help.

- **Potatoes.** Root vegetables are great since they require little assistance and can grow in less than ideal soil.

- **Sweet Potatoes.** I personally prefer these over regular old russets, both for taste and nutrition since it has more vitamins. The difference is minimal here though for survival purposes.

- **Carrots.** Another root vegetable with lots of good vitamins.

- **Spinach.** Great nutrients, but calories are minimal. You need both to survive long term, but a couple leaves of spinach per week is plenty, so ratio your crops accordingly.

- **Beans.** There are dozens of varieties, but they are generally all packed with energy and have some protein too. That's a big bonus.

- **Garlic.** I know I said it wasn't about being gourmet, but garlic takes the most minimal of space, you don't need a whole lot of it, and it has some important health properties such as being anti-inflammatory and anti-viral. Plus it tastes great.

- **Cabbage:** easy to grow, decent energy and calories.

- **Herbs:** just for flavor for the most part, but like garlic, it takes almost no space to grow some rosemary, basil, thyme, and onions.

- **Winter squashes:** like the name implies, they store excellently during the winter months. This includes acorn, spaghetti, and butternut squashes, as well as pumpkins. As long as you don't cut into them, they can store for months in a cool place.

If you've never grown anything before, don't wait until you're close to starving to test out your green thumb. Go to the garden store, get some seeds, and set up a little garden, it can seriously be just a big pot or two on your porch. Do some trial and error and figure out what works for you and what doesn't. You don't have to be so committed you convert your entire yard into a farm, but at least have an idea of what you're doing.

Quick tip for a speedier harvest: go ahead and allow everyone to urinate in the garden. It's water plus nutrients, a win/win scenario for growth. Avoid using "human fertilizer" as there are too many nasty bacteria and gross stuff that require some type of treatment first to be safe, and you don't have that kind of time or materials. If you don't get the polite reference, I'm saying don't poop in your garden.

Digging your garden can prove to be a challenge. If you have a shovel, fantastic! I recommend using that! The chances are that a shovel isn't one of the lightweight "must-haves" that someone in your survival group thought to grab, although it could double as a weapon. Damn, now I really hope someone has a shovel. No shovel, no worries, there are plenty of ways to dig without one. You can use just about anything including but not limited to your hands, sticks, sharp stones, or explosives (you've gotten to know me pretty well by now, do you think that's something I seriously recommend?).

One quick note about excessive digging or gardening... It is just as important to protect your hands as it is to protect your feet. Take turns with the other members in your survivor group. It's far better to have everyone's hands a little sore and blistered than it is to have one party member (especially if that member is you) completely incapacitated.

A final thought is protecting your crops from thieves. Think about how you can both secure and defend your crops. A big fence with razor wire on top would be ideal, but if nothing else, keeping the garden in a green-house with a lock on the door is better than nothing as it will stop the casual thief. It keeps honest people honest.

People take a lot of pride in growing their own food. The Zombie Apocalypse in not a time for pride. It is a time for emotionless essentials and black and white basics. You must be ready to abandon your garden in an instant's notice. Don't spend all day pulling weeds and making sure everything looks good. Utilize your time by harvesting what can be eaten immediately, which seeds can be packed up, and what you can preserve for later.

If you get too attached to your food source and temporarily permanent housing, you may want to return to it should you be forced out. This is ill-advised. If humans force you out, they are likely stronger and better prepared than you in the first place. Move on and start over. If zombies force you out, they could have contaminated all food sources and if they can get to you once, they can get to you again. If nature, weather, or animals force you out, then that means your shelter sucks. You should definitely move to a better location and try harder while securing this one.

Hunting food

If you've never hunted before, there is a reason it is called "hunting" and not "killing". So, if you fit in that group of "I haven't hunted before", now is the time to start practicing. Even the best hunters and fisherman in the world get skunked sometimes, which really sucks if you haven't eaten in a few days. Don't wait until the zombies have taken over to figure this one out.

Personally, I'm not living through an apocalypse just to eat lettuce the rest of my life. Vegetables are what my food eats. You've got two choices to be a carnivore here: raise it or hunt it. Luckily, it's become socially acceptable to raise chickens in the city, not just the country, so you may be pleasantly surprised how much fowl is on the loose. A few chickens can provide protein via eggs of course, but you may not want to eat the chickens unless you have a rooster that can help replace them. Problem is that municipalities don't always allow roosters because they are a nuisance for their loud noise so it's off to the farmlands if you feel need to breed. Generally, this isn't a bad idea since by definition it will be further away from dense populations of people (and people become zombies).

Cows and goats provide milk and meat while pigs, lambs, and turkeys are usually worthwhile to raise and butcher. You always have to weigh the effort and risk vs reward though, just like in any other survival situation. These animals need constant care and food. Are they eating grass, grains, or corn? Also, the better looking your situation, the more it's going to attract attention.

Hunting is a great option, but you have to consider the ROI of expending bullets or shells. Carrying around a bunch of #6 birdshot just to kill a few doves is a waste of time and energy. You get maybe a single bite of dove meat, maybe two small bites if it's a big one. As far as birds and fowl go, I'd stick with ducks and geese and larger birds. Do whatever you can to make sure you take the animal cleanly and in one shot so as not to waste good meat or valuable ammo. Yup, that's right. I'm telling you to shoot a goose in the face. Head shots kill and don't waste meat. Tough times call for tough men, women, and children. And unfortunately, for tough meat.

When it comes to large game, a single well-placed bullet can yield dozens of pounds of meat. With deer, I would suggest baiting them with dried corn if you have it and taking an easy shot. You are not hunting for sport here. Consider them like you would a cow or pig that you are slaughtering for food. Make the kill clean and swift, you don't need to waste energy tracking and dragging a deer or wild hog back home. Also, don't forget that every time you go out tracking an animal, you leave a trail that will allow somebody, or something, to track you.

Preserving Food

Remember the fable of the lazy grasshopper who played all summer while the ants collected food? Don't be the grasshopper, be the ants, or else you will find yourself eating grasshoppers and ants and that's if you're lucky! When winter comes, even in warmer climates, crops will stop growing or get killed by the cold, so you'll need to have planned ahead. If you are out of rice and beans and pasta, having canned food from your garden is a big win. Having no food from your non-planning is a non-victory, also known as a loss.

Canning is relatively easy in concept: you put food in a jar, partially submerge in boiling water, boil for a prescribed amount of time, then seal and store. However, if you have not actually done it before, it would behoove you to try it out before any apocalypse.

I'm not writing a book about canning, but lucky for you, there are literal thousands of videos and millions of webpages online if you're looking for a short read. If you're serious about it, there is a plethora of books written by experts devoted to the subject. I know so few of you are going to figure out how to can food pre-Apocalypse, but do me a favor. Search your brain and see how much useless information you have in it. Do you watch 20 plus hours of sports each week? What about reality TV?

There's a difference between down time and dead time. Let me explain. Down time is when you are in between tasks, whether it's evenings at home after your 9-5, a long weekend thanks to Labor Day, or a long lunch break. Down time can be used to accomplish things like pay bills, mow the grass, read a book, exercise, do laundry, etc.

Dead time, on the other hand, is down time where you decide to quit being a functioning human being. This includes your drinking benders for those of us age 21+, post pancake naps, binge-watching TV with no educational value, play Call of Duty (hey, that one's negotiable), or anything else that requires absolutely no effort and, consequently, no brain power.

I for one believe that dead-time is necessary and healthy, when done in moderation. We all deserve those brief moments of bliss where we can escape from the demands of a chaotic life. However, you should have way more down time than dead time. In fact, I can guarantee you have way too much dead time in your life.

Bottom line, the point of this point is simply this: You have time to learn how to can food.

A different problem presents itself with large game. You can take what you need, as you need it from the garden, but once a deer is killed the clock is ticking. There are a few basic options. First, you can certainly can meat just like a vegetable. If you want, make your own venison stew with your vegetables and then can it for a ready to eat meal later, same thing as Campbell's or Dinty Moore. Chicken and fish are great for canning as well. Again, I'm not writing a canning book, so you'll need to do your research and practice. Next, you can salt, cure, or dry many different meats including venison, fowl (chicken, turkey, etc), and even meatier fish like salmon. The simple process is you slice the meat very thin, brine it or heavily salt the meat, cook it at a low temperature (200 degrees or so) for around 4 hours or more depending on temperature, then let it dry out in the oven or smoker. You don't have to smoke it, but smoke is actually a good preservative too, and it tastes good. After that, the meat can be stored in a dry container for weeks or months. You don't have to eat it like a snack, in fact, you might want to rehydrate a little to get some of the excess salt out.

FOOD

What would be really ideal is if you are in a good strong group of survivors though. For a variety of reasons of course, but when it comes to splitting up the work, food is a major priority. Growing, hunting, and preserving food is a LOT of work and time. What would be real nice is to have fresh meat and vegetables and fruit every day because you take turns hunting and when you take down a big deer, everyone can eat from it for a day or so.

Cooking Food

This one I can sum up for you very quickly. The luxury of enjoying a juice prime-grade steak so rare a good veterinarian could probably bring it back to life is gone from your life. From this point forward, all meat is to be cooked well done. It wouldn't hurt to cook all of your vegetables thoroughly either. If there is a contaminant in blood that is causing humans to turn into flesh-craving zombies, then there's no telling what is in the land, air, and water. Zombie-factor aside, you really don't want to risk getting the Hershey-squirts from undercooked food when there is no hospital or medicine available.

You may or may not want to keep a fire burning the entire time depending on outside temperature, nearby kindling and lumber, and if you are staying hidden or trying to attract other survivors. You don't want to build three fires to cook three meals. Build one fire and cook all three meals for the day (that is if you're one of the few survivors living like a king) or keep one burning.

TLDR:

Be sure to keep a regular stock of store-bought cans of vegetables, rice, pasta, and beans. This can hold you over for a week or more easily. If you have any intention of staying put, invest in some dehydrated meals to store away in the basement. That food will eventually run out, so plan on stretching it by growing your own crops and hunting food. Learn to replace your own long-term food stores by canning and preserving food. YOU need to learn how to do this stuff, so practice now: that includes the growing, hunting, canning, and preserving.

Water

When the well's dry, we know the worth of water.

- BENJAMIN FRANKLIN

We all know how important water is to our survival, but many people discount some important factors. If you ask someone to show you how much water they'd need for a week, the average person would be woefully underprepared. Further, many folks don't realize how heavy water is, which is a wildly important factor if you are traveling. The standard we have all heard our entire lives from doctors, fitness gurus, and skinny celebrities on television is "8 glasses of water per day." I recommend more, especially if you are traveling a lot and fighting, but let's set that as our golden standard. That's about 2 liters, or ½ of a gallon. You can go for a long period of time with much less but try to avoid that as it is terrible for your body, especially long-term. *The point being, that much water is 3.5 gallons per week, which is about 30 pounds of water.* Trust me, you don't have the room in your pack or the strength to lug that much water plus all your other gear. And that's just water for drinking! There's also hygiene, first-aid, farming, food prep, and area cleanliness to consider. All of which are wildly important factors.

Water for a defensible location

If you have a defensible location, you have to consider your resources as something that has to be protected. The same amount of manpower and energy spent building your set up, must be spent defending it. Depending on the size of your camp, I would have at least four armed men guarding the perimeter, keeping watch in each direction.

Wherever you are, is where your water should be. If you have a nice bunker, but your water is outside, you won't last long if you get attacked. That means don't keep your 50-gallon barrels outside where you can't protect them and can't get to them. Sure, you can keep some "extra" barrels outside and out of the way, especially those used for collecting water, but keep your main supply hidden, secured, and, most importantly, with you.

Collecting Water

As for collecting water, the most basic concept is to catch the run-off from your roof into a barrel. There are plenty of other ways and I recommend using any and all of them. It will never hurt you to have too much water so the more systems you set up, the better. Keep in mind there is no weather channel, app, or local meteorologist to piss you off with inaccurate predictions. You never know when the rain is going to come and when it does you must capitalize on the opportunity. So, dig a hole, spread out your tarps, and stand outside in the rain with your mouth open. The only wrong way to gather water is to not gather water.

There are a lot of natural filtration systems and a lot of ways to set up contraptions that will filter water for you as it is collected. Here is my opinion on those:

In times of war or famine, natural springs or man-made devices that offer drinkable water are a godsend and an absolute survival must. In times of a zombie apocalypse, I'm not drinking anything I haven't sterilized. Not only am I not drinking anything that hasn't been boiled, I'm not cooking with, washing with, cleaning with, swimming in (if avoidable), or anything else that I haven't personally sterilized. Do you know what the Government is capable of contaminating the earth with that would cause an Apocalypse of this nature?! Yeah, me neither. So, always boil up before bottoms up.

Natural Water Source

What if you set up camp near a water source like a river? As long as it's not perilous to reach it such as having to navigate through territory that makes you vulnerable to ambush or simply that the water supply could go dry or go bad, that's a great option. It's such a great option, that it's extremely likely it's the option that many survivors will think of it. This means you will either have many groups of survivors in your nearby vicinity which increases your likelihood of being discovered by zombies or you will have somebody waiting to steal your setup.

How would I ambush someone with a nice-looking camp near a water source? I would wait until a few of them separated from the big group to get water from the source. Then, what are you going to do when members of your water retrieval party don't come back? Don't you send a few other members to look for them (I'm killing them instantly) or a large group to look (I'm taking over your camp). Relax, being near a water source is worth the risk and it's likely that not all humans you will come across are as savage as me. Maybe y'all will party it up on the river and get along like strangers on spring break. I'm either a pessimistic realist or a realistic pessimist so take what I say with a grain of salt. Which reminds me, salt dehydrates you. Watch your salt intake or increase your water intake, especially if you're chowing down on preserved meats.

A better option would be to set up an irrigation system that brings water from the river to you. In this case you will want to damn the river and build a series of trenches that force the water to come to you. You don't need a massive dam here guys; I'm talking just enough to subtly divert a tiny stream that gives only as much as you need. You can add different grasses and filtration systems to ensure the water is cleaned up and silt is filtered down before it reaches you or your crops, but you know how I feel. The only safe water is sterilized water. If you do set up an irrigation system, be sure to have a gate or other type of defense that extends to below the trenches entering your camp. Every cartoon dog knows the best way out of a fence is under a fence and if you've already dug trenches, you've given the enemy a head start.

TLDR: Personally, I'd try to have at least 50 gallons per person on hand. That does not mean you drink from the 50 gallons, let it get down to 5, then go replace it every week. Rather, you have that 50 gallons as your reserve in case the crap hits the fan and you cannot get to your water supply for an extended period of time. It's just like making sure your bank account doesn't go under $1000, because you know that life is bound to throw $1000 worth of surprise at you during any given time – typically as soon as you get comfortable and happy. The Apocalypse won't be any different. Oh wait, yes it will. It will be way worse. There will be more surprise garbage then you can fill the ocean with.

First-Aid and Medical

"Laughter is the best medicine, except for treating diarrhea."

So, what's "basic" gear a SEAL would choose for medical supplies? You might be surprised. If I had nothing on me and ran into a grocery store or drug store, not a specialty store with mountain survival kits hanging on the wall, first and foremost: **Ibuprofen, Aspirin, and Tylenol.** If you can, antibiotics too.

Yup, you read that right. Not bandages, not splints, not tourniquets, not an AED, or any other fancy gizmo. That is not BASIC stuff, and people have a tendency to discount and forget the basics. You can do a lot with anti-inflammatory and pain/fever medications. You can try to get some antibiotics but if you don't know what you're doing with them, you might cause more damage than you fix (caveat if it's life or death, you might have to roll the dice).

If you anticipate being outside, maybe some sunscreen, but it's going to be a premium luxury item. At the end of the day, skin cancer takes a while. Dehydration, starvation, and a thousand other things will kill

First-Aid and Medical

you a lot quicker, but it can make life less miserable and help you move faster. Depending on what kind of environment you're in, bug spray can avoid a miserable night with all the mosquitoes buzzing, just for your mental state of getting some rest. Then again, that's a consumable product and when it's gone, it's gone, so you better have a better long-term solution before then to keep the bugs off.

Last but not least: ***duct tape and zip ties***, because they're easy to pack, they're light, and useful for a whole laundry list of things. You can always find sticks to make a splint for a broken limb, but duct-tape secures it. You don't want to bring a whole bunch of bandages, because you are literally wearing bandages in the form of your clothes. If you've got some great, but not a basic necessity.

I know my first-aid, but I'm also not arrogant or stupid. One of my buddies is Dr. J.C.H. (partially redacted, like me). He is a retired Army Colonel in the Medical Corps who was attached to the 75th Special Operations group back in the day. As a trauma surgeon with over 3 decades experience, he's put more broken bodies back together from unspeakable injuries than you've had hot meals, so pay attention:

While cures have been discovered for ailments such as tuberculosis, trauma will remain a leading cause of death for as long as humans exist. In the United States, in the age group of 45 and younger, trauma is the leading cause of death. In persons under 30 years of age trauma accounts for more deaths than all other causes of death combined. While a great number of trauma deaths can be prevented in our society (car crashes involving alcohol, failure to use seat-belts, etc.) trauma deaths are varied in their underlying mechanisms and may be related to many forms: falls, injury from blunt objects, stab/impalement, etc.

In any circumstance (post-apocalyptic or otherwise) it is vitally important to identify the 3 main categories of trauma related deaths:

1. **IMMEDIATE:** Approximately 60% of deaths from civilian trauma occur simultaneously with the traumatic event. That is, the victim dies almost instantly. Most are due to mortal brain, heart or major vascular injury or airway or ventilation compromise. Short of the trauma event happening on the operating table, nothing can be done. Even then not much can be done for mortal brain injuries.

2. **EARLY:** Another 20% of trauma deaths occur within the first few hours of the event. It is in this group that initial management of the victim at the scene of the trauma can make a difference in terms of both morbidity and mortality (ie – "morbidity" means being afflicted with a disease or medical condition, "mortality" meaning you died from it or will die from it). Most common in this subset are instances where major bleeding can be stemmed, an obstructed airway cleared, breathing assisted using a ventilator, and CPR used to restore circulation. This is the level where basic first-aid training and experience is needed and can be the difference between quick recovery and death.

3. **LATE:** These deaths occur in the days or even weeks following the event, and after care has been provided. These are deaths that are due to late infections and sepsis, ventilatory failure, heart attacks, renal insufficiency, "multi-system organ failure", and the like. Immediate assessment, treatment, and cleaning of wounds in the first few minutes or hours (ie – EARLY stage) decreases the likelihood of late trauma death.

When 911 becomes ancient history, any individual who is onsite at the trauma event is by definition a "first responder" and has the opportunity to improve the victim's outcome in that immediate phase of the scenario. Whether that person is adequately trained or not is another question.

First-Aid and Medical

Knowing the techniques of first-aid is important, but the onsite care of the trauma victim must follow certain guidelines, the cardinal of which is *"Deal first with issues which are immediately life-threatening"*. This seems obvious, but it is surprisingly misinterpreted and easily forgotten when stress and fear are factored into the equation.

First, the victim must be withdrawn from any source of ongoing injury. This is not merely pulling the victim from a burning fire or repeated blows from a blunt object. Rather, it is the first-responder not succumbing to the sight of gruesome injuries to the face, missing limbs, open or "compound" fractures (the bone is sticking out from skin and muscle), severe dislocations of major joints (limbs are twisted at unnatural angles), open abdominal injuries (you can see guts and organs), and the like which can psychologically overwhelm the inexperienced first-responder. These visually dramatic injuries do not trump the importance of maintaining an airway, breathing, and circulation. The "immediately life-threatening issues" are often non-visible situations which, if left unattended, result in death or permanent disability within scant minutes. Remember the ABC's:

AIRWAY

- First, is anything blocking the airway? Facial fractures, food or other foreign bodies in the mouth or throat, a tongue "swallowed" into the oropharynx, all can be managed with fairly simple measures like a "jaw thrust", tongue traction, or simply removing the foreign material. In layman's terms, pull out whatever is choking the person. HOWEVER, and this is an extremely important "HOWEVER", do not ever do a "blind sweep" which is just jabbing your finger in the victim's mouth without seeing what you're doing because:

- if you don't see anything choking the person, or don't know that they are choking, you could hurt the person by digging around in their mouth and throat.
- if they are not actually choking on anything, you waste precious time doing nothing helpful.
- You risk lodging a foreign body further into their airway causing more obstruction

BREATHING

- If the airway is clear, they may not be able to breath on their own. You need to perform CPR. I know this was referenced in an earlier chapter, but it bears repeating and some additional details.

- For pediatrics with 1 rescuer- 2 breaths for every 30 compressions.

- For pediatrics with 2 rescuers- 2 breaths for every 15 compressions (one person breaths, one person does compressions).

- For adults the ratio stays 2 breaths for 30 compressions with 1 and 2 rescuers

- Conduct chest compressions at a rate of 100/120 per minute. That means 2 compressions per second.
  - Trust me, you won't be able to keep count in the heat of the moment. So, be like the Bee-Gees, and complete chest compressions to the rhythm and speed of "Staying Alive".
  - "I – I – I - I'm…staying alive…staying alive." Yes, it is 100% OK to sing out loud, even if you don't sing on key. Yes, I'm absolutely serious. The right pace is imperative for CPR to be effective, and you will be stressed beyond belief. Objectively, you should sing it out loud to keep yourself focused.

- It is worth mentioning that CPR is extremely taxing on the person doing it. It may not seem like it's that hard of an exercise, but it

First-Aid and Medical

can wear you out in just a few minutes. If you have someone else with you, switch every 5 cycles or about 2 minutes. If you are by yourself, the reality is you just have to try your best and keep going until you can't go anymore.

- If the first responder is sufficiently experienced (not just trained) to recognize an otherwise fatal "tension pneumothorax", a life-saving small intentional stab wound can be made on the effected side of the chest, sustained by insertion of a small rigid tube (e.g., the empty body of a ballpoint pen or a straw). An "open pneumothorax" with positive pressure ventilation is survivable whereas a tension pneumothorax is not.

CIRCULATION

- What does this entail exactly? Getting the blood in the body where it needs to go- important/vital organs such as your brain

- Even if you aren't 100% up to date on your CPR training, don't hem and haw, just doing "Hands-only CPR" is better than nothing, and an imperfect rhythm of CPR is still better chance than doing nothing. However, performing aggressive CPR on a victim with no airway or ventilation may as well be CPR performed on a training dummy.

Hemorrhage/Bleeding

Hemorrhage into the environment, even when pulsatile or startling in volume (that is, the victim is spraying blood out of an artery and shooting it across the room), can usually be stemmed by applying direct pressure to the bleeding site or compression of a major arterial pulse above the level of the bleed. "Above" means closer to the heart than where the bleed is. Examples: elbow, armpit, knee, thigh/groin, etc.

Application of an effective tourniquet is a last resort measure and unless judiciously released/reapplied at 15-20 minute intervals can usually result in more damage than good. An exception here would be a tourniquet placed immediately above the site of a traumatic limb amputation such as a missing hand, foot, or even arm/leg below a joint. Unfortunately, little if anything can be done by the average first responder in the case of major hemorrhage arising from the interior of the chest, abdomen, or pelvis. In layman's terms, if the person is gut-shot or stabbed and bleeding inside their body where you can't see it, you can't do much in the field. Rapid transport to a definitive care facility is likely the only hope as a surgeon will need to open them up to stop the damage.

The same can be said for injuries to the brain and central nervous system where little can be done in the field, however measures must be taken to prevent worsening of existing injury. For example, a victim might appear to be conscious and neurologically unimpaired, but as a first responder you should assume that injuries to the spine and spinal cord area are present and account for that when positioning and transferring of the injured are performed. In other words, don't move the victim's neck or spine any more than you have to (preferably not at all).

In cases involving penetrating injuries by such objects as knives, branches, or similar sharp devices (particularly to the torso, head, and neck), the inclination is to immediately remove the object. DO NOT. You must suppress that instinct, especially if the penetrating object is seen or felt to transmit pulsations, meaning the object is moving as the person's blood pulse flows by. Get the person somewhere stable and try to find experienced help if at all possible. It's counter-intuitive to leave a knife or other object lodged in someone but pulling it out could mean the victim bleeds out within seconds.

First-Aid and Medical

A cool head must prevail when dealing with trauma at the scene. Follow these guidelines:

1. Evaluate the circumstances before taking action
2. protect the victim from ongoing injury
3. seek help if at all possible &provide exact information to responders
4. assume the worst when anticipating the extent of injuries
5. deal first with immediately life-threatening injuries
6. remembered your ABCs (Airway, Breathing, Circulation)
7. stop bleeding
8. assume damage to the central nervous system (especially cervical spinal injuries)
9. reduce and splint long bone fractures
10. do no harm! How? Stay calm, cycle back to step 1.

Know your Zombies

"In those moments where you're not quite sure if the undead are really dead, dead, don't get all stingy with your bullets."

– ZOMBIELAND

A critical aspect to surviving the zombie apocalypse is to determine what flavor of zombie you'll be dealing with. As I see it, there are two main categories. You've got your fast-movers and slow-movers.

Slow-Movers

Let's use the AMC's "Walking Dead" series as an example. These guys are *nasty* in a way that defies any logical explanation. The general modern-day science-y concept of zombies to my knowledge is that virus kills the host, and once dead, causes the nervous system to start firing artificially with the singular goal of eating other living things. To an extent, this makes sense, as we can assume it is the same as any virus's basic purpose, to make more of itself. This is all good and well, and a fine explanation…except that half these things are little more

\\ Know your Zombies

than skin and bones and all their tissue has decomposed and rotted, meaning there is no way for the nervous system to be intact.

That's boring science though. I don't' really give a crap why zombies act like zombies. I just don't want to be one. Here's the strengths and weaknesses you need to concern yourself with.

Weaknesses:

- **Slow AF.** Seems like it doesn't matter if they just turned into a zombie or are little more than bones and sinew, they lumber around like they are a mildly brain-injured cat with laryngitis stuck in wet concrete.
- **Super noisy.** For some reason, they can't stop themselves from being irritatingly loud with their groaning and moaning (Like your mom. BURN!).
- **Not very smart.** You'll see these dumbasses stuck in little ditches or unable to navigate around a parked car. They just keep bumping into the thing like a Roomba that went haywire. They literally get themselves impaled on inanimate objects. This works to your advantage because you can just set out a sharp stick at about 4 feet high, and they'll just run into it. Then again, it's a disgusting job to dislodge them from it.

Strengths:

- **Super sneaky.** So, while they are not smart, not fast, and super loud most of the time, they somehow manage to sneak up on you in the middle of a deserted WalMart parking lot and bite you in the neck. Those SOB's just inexplicably come out of nowhere about 5% of the time, so don't let your guard down.
- **Brain only shots.** They can be missing limbs or *their entire body*, and those butt-holes will STILL try biting you! It doesn't even

make sense, they don't have a stomach attached! Aaaanywho, the only way to stop them is putting a foreign object into their brain box. Unlike on Walking Dead, don't try stabbing someone in the temple with a knife. Believe it or not, the human skull is pretty hard and you won't punch through it with a pocket knife. The eye works best for a knife…nasty, but I'm just saying it works.

What to do?

Just be careful they don't sneak up on you with their voodoo quiet magic, but otherwise, take your time, carefully place your shots, and conserve ammo. You've got more than enough time to line up your sights. "one shot, one kill" is the motto here.

Fast-Movers

Let's get jiggy with it. The adaptation of "I Am Legend" starring Will Smith is a perfect example of a zombie outbreak that probably involved a meth-lab. These dudes can MOVE. They have some serious speed and strength, including climbing up a 2-story house in about 5 seconds and punching their way through the roof and attic. Here's something interesting though, that is actually based in science. Have you ever heard how a mom can lift a car off their child? It's true. In terms of pure physics, removing any neurological notion of pain or cognizance, the human body is capable of far more than we think possible. There are numerous stories such as a mountain climber bench pressing a 700 lbs rock off their chests. It's possible, but it will rip the ligaments and tendons and muscle fibers. Point being, it's possible.

But what if there was a half-way point, where pain is removed from the equation, but a sense of self-preservation remains. Allow me to explain: let's say an average 180 lbs male in his 40's gets bit and turns into a zombie. As a human, this guy can bench press around 200 lbs. The physics of his body however would allow a maximum of about 800 lbs, but again, would result in catastrophic pain and damage to his body. But if pain was a non-factor, and say he limited it to 400 lbs which would not *overly* damage his body, that's your fast-mover.

"Quit with the science lecture!" You say, "this is a zombie book! Get to the point!" you say. OK, the point is these zombies are faster and stronger than you. That is a really bad problem.

Weaknesses:

- **No sunlight.** This is your one saving grace. They are fast as hell and meaner than a snake, but no SPF can help them with the sun. Plan your days to do what you need to do, then get the hell inside a secured area. Also, some tanning bed lights could be a potential weapon, I guess.
- **Pretty insane.** While fast and seemingly organized, they have zero regard for their own well-being in terms of a team of some sorts. They'll sacrifice themselves for the good of the group. They clamber on top of each other, jump in front of cars, etc. You can use that to your advantage if you plan ahead.
- **Susceptible to non-brain related injury.** While it may not put them down permanently, they don't like being shot in the gut or extremities, and seems to slow them down.

Strengths:

- **Very, very fast.** They run at Olympic sprinter speeds, so don't even think about outrunning them. You'll have to stand and fight.
- **Organized.** They seem to go in groups, but not like slow-movers. The slow-movers just wander aimlessly like sheep in a herd. The fast-movers seem to have a target in mind, although I wouldn't go

so far as to say they have a "plan". That relates back to their weakness of being insane and having disregard for themselves.

What to do?

Center of mass, baby. That means it will require 2-3 rounds on average, assuming you don't suck at shooting, aimed at the middle of their chest. This will stop them or at least slow them down considerably. Once slowed down, you can finish them off with a dome shot.

Final Thoughts

If you've made it this far, thanks for reading. I hope you enjoyed it and picked up at least a few helpful tidbits along the way. Like I hope I made clear, this book isn't the "how to build a mud-hut" and live in the wild for decades, although I do reference some aspects of that.

I felt like there was a gap in survival books for a practical, pragmatic, and realistic look at a medium to long term to survival situation. The idea being that things could and would return to normal, but it could be a weeks, months, or years. "Weeks" may not sound that bad, but remember what happened when COVID first hit and people panicked and cleared out grocery stores in a matter of hours or days? That wasn't even a real emergency, but highlights the lack of planning by most people, and how quickly things can go sideways.

I try not to take anything too seriously, so hopefully my attempts at humor and injecting a little sarcasm didn't fall totally flat. There is a practical reason for it though, and that's that making a book more engaging helps the important stuff stick in your brain better. I hope that's the case here.

\\ **Final Thoughts**

Jokes aside though, thanks again for reading and pat yourself on the back because as you'll see on the next page, your buying this book made a difference in someone's life.

The Navy SEAL Foundation

In lieu of taking royalties, the author has directed that those funds be given 100% to the Navy Seal Foundation. By purchasing this book, you've helped make a difference in the lives of the families of America's elite warriors who have fallen in combat and SEALs who sustained injury in duty, sacrificing opportunities and the ability to work or live as they otherwise would have.

Incredibly, 94 cents of every $1 donated is provided directly towards their mission objectives, meaning it is a very lean and effective operation. Your money is helping where it is needed, not being wasted or lining anyone's pockets.

In other words, you can feel pretty darn good about this!

Check out more info at NavySEALfoundation.org

www.ingramcontent.com/pod-product-compliance
Lightning Source LLC
Chambersburg PA
CBHW070116080526
44586CB00013B/1311